AF522083

PHYSICAL GEOGRAPHY

(ATMOSPHERE)

PHYSICAL GEOGRAPHY
(ATMOSPHERE)

By

Dr. K. Bharatdwaj

DISCOVERY PUBLISHING HOUSE
NEW DELHI-110002

Reprinted - 2019

First Published - 2006

ISBN: 978-81-8356-151-8

Physical Geography
(Atmosphere)

Published by:
DISCOVERY PUBLISHING HOUSE PVT. LTD.
4383/4B, Ansari Road, Darya Ganj
New Delhi-110 002 (India)
Phone: +91-11-23279245, 23253475; 43596065
E-mail: discoverybooksindia@gmail.com
discoverypublishinghouse@gmail.com
web: www.discoverypublishinggroup.com

Printed at:
Infinity Imaging Systems
Delhi

Preface

It may well be said that there can be no geography which concerns itself with the actual shape and form of the land surface, solid rode, the configuration and extent of the seas and oceans, the enveloping atmosphere without which life as we know it cannot exist, the physical process which take place in that atmosphere. This book has been designed to cover the syllabus of physical geography required for the B.A. students of the Indian Universities. The subject matter has been arranged so as to provide clear and integrated approach to the subject with all essential tools of applicable geography for B.A. curriculum.

Care has been taken to make the treatment of the subject simple and accessible to the average students. It is believed that the book in present form will be found to be useful by the student community and the teaching fraternity alike.

Suggestion for the improvement of the book will always be most welcome.

Author

CONTENTS

G.T. Trewartha, Equatorial Climate or Tropical Rain Forest Climate (AF), Monsoon Climate (AM), Savanna Climate (AW), Tropical-Subtropical Hot Desert Climate (BWH) (Sahara Type of Climate), Mediterranean Climate (CS Climate), China Type of Climate (CA), Steppe Climate (BSK), West European Type of Climate (CB), Boreal or Subarctic or Taiga Type of Climate, General Features of Climates, Climatology and Meteorology, 'Local Climate' and Micro-climatology, Climatic Factors and Elements, The Atmosphere, Tundra Climate.

1

COMPOSITION AND STRUCTURE OF THE ATMOSPHERE

COMPOSITION OF THE ATMOSPHERE

The atmosphere is a thick gaseous envelope which surrounds the earth from all sides and is attached to the earth's surface by gravitational force. The atmosphere is a significant component of the biospheric ecosystem because the life on the earth's surface is because of this atmosphere otherwise the earth would have become barren like moon. Besides providing all necessary gases for the sustenance of all life forms in the biosphere, it also filters the incoming solar radiation and thus prevents the ultra-violet solar radiation waves to reach the earth's surface and thus protects the earth from becoming too hot. The height of the atmosphere is estimated between 16 to 29 thousand kilometres from the sea level. It is estimated that 97 per cent of the effective atmosphere is upto the height of 29 km. In fact, the air is mechanical mixture of several gases. The atmosphere is composed of :

(i) Gases,

(ii) Vapour, and

(iii) Particulates.

Gases

Nitrogen (78%) and oxygen (21%) are major gases which constitute 99% of the total gaseous composition of the atmosphere. The remaining one per cent is represented by argan (0.93 %), carbon dioxide (0.03%), neon (0.0018%), helium (0.0005%), ozone (0.00006%), hydrogen (0.00005%), krypton (trace), xenon (trace), methane (trace) etc. Oxygen is the most important gas from the stand point of living organisms because they inhale it for their survival. Oxygen is also essential for combustion of burning matter. Nitrogen acts as dilutent and is generally chemically inactive. Carbon dioxide is used by green plants for photosynthesis. It absorbs most of radiant energy from the earth and reradiates it back to the earth. Thus, carbon dioxide, a green-house gas, increases the temperature of the lower atmosphere and the earth's surface. The concentration of carbon dioxide in the atmosphere is gradually increasing due to burning of fossil fuels (coal, petroleum and natural gas) and deforestation. Ozone gas absorbs most of the ultra-violet rays radiated from the sun and thus prevents the earth from becoming too hot.

Water Vapour

The vapour content in the atmosphere ranges between zero and 5 per cent by volume. Climatically, water vapour is very important constituent of the atmosphere. The atmospheric vapour is received through the evaporation of moisture and water from the water bodies (like seas and oceans, lakes, tanks and ponds, rivers etc.), vegetation and soil covers. Vapour depends on temperature and therefore it decreases from the equator poleward in response to decreasing temperature towards the poles. The content of vapour in the surface air in the moist tropical areas, at 50° and 70° latitudes is 2.6%, 0.9% and 0.2% (by volume) respectively. The content of vapour decreases upward. More than 90 per cent of the total atmospheric vapour is found upto the height of 5 km. If there is condensation of all the atmospheric vapour at a time, there would result a one-inch thick layer of water around the earth. Even this meagre amount of water vapour in the atmosphere is responsible for various types of weather phenomena. The moisture content in the atmosphere creates several forms of condensation and precipitation *e.g.* clouds, fogs, dew, rainfall, frost, hailstorm, ice, snowfall etc. Vapour is almost transparent for incoming shortwave solar radiation so that the electromagnetic radiation waves reach the earth's surface without much obstacles but vapour is less transparent for outgoing shortwave terrestrial radiation and therefore it helps in heating the earth's surface and lower portion of the atmosphere because it absorbs terrestrial radiation.

Particulate Matter

The solid particles present in the atmosphere include dust particles, salt particles, pollen, smoke and soot, volcanic ashes etc. Most of the solid particles are kept in suspension in the atmosphere. These particulates help in the scattering of solar radiation which adds varied charming colour of red and orange at sunrise and sunset. The sky appears blue in colour due to selective scattering of solar radiation by dust particles. Salt particles become hygroscopic nuclei and thus help in the formation of water drops, clouds and various forms of condensation and precipitation.

STRUCTURE OF THE ATMOSPHERE

The modern knowledge about the atmosphere is based on the information received through rockets, radar and satellites. The effective height of the atmosphere is estimated between 16 and 29 thousand kilometres from the sea level but the height of the atmosphere upto 800 km is most important.

About 50 per cent of the atmosphere lies below the altitude of 5.6 km and 97 percent of the atmosphere is confined to the height of only 29 km. The upper limit of the atmosphere, though unknown, is considered to be 10,000 km from sea level. The earth's atmosphere consists of a few zones or layers like spherical shells. On the basis of the characteristics of temperature and air pressure there are four layers from the earth's surface upward *e.g.* :

(1) Troposphere,

(2) Stratosphere,

(3) Mesosphere, and

(4) Thermosphere

(Fig. 1.1).

(1) Troposphere

The lower most layer of the atmosphere is known as troposphere and is the most important layer because almost all of the weather phenomena (*e.g.* fog, cloud, dew, frost, rainfall, hailstorm, storms, cloud-thunder, lightning etc.) occur in this layer. Thus, the troposphere is of utmost significance for all the life forms including man in the biospheric ecosystem because these are concentrated in the lowest part of the atmosphere. Temperature decreases with increasing height at the rate of 6.5°C per 1000m. This rate of decrease of temperature is called normal lapse rate. There is seasonal variation in the height of troposphere. In other words, the height of troposphere changes from equator towards the poles (decreases) and from one season of a year to other season (increases during summer while it decreases during winter). The average height of the troposphere is about 16 km over the equator and 6 km over the poles. The upper limit of the troposphere is called *tropopause* which is about 1.5 km thick. The height of tropopause is 17 km over the equator and 9 to 10 km over the poles. There is also seasonal variation in the height of tropopause. Its height is 17 km during January and July over the equator and the temperature at this height is –70°C. The height of tropopause during July and January over 45°N latitude is 15 km (temperature –60°C) and 12.5 km (temperature –58°C) respectively. The height decreases further poleward as it is 10 km during July (temperature –45°C) and 9 km during January (temperature –58°) over the north pole. It is apparent that temperature at the top of tropopause is lowest over the equator (–70°C) and is relatively high over the poles. Since temperature decreases upward at the rate of 6.5°C per 1000m and hence it is natural that temperature at the height of 17 km over the equator becomes much lower than at the height of 9-10 km over the poles. The word troposphere literally means 'zone or region of mixing' whereas the word tropopause means 'where the mixing stops'.

(2) Stratosphere

The layer just above the troposphere is called *stratosphere* but there is contrasting opinion about the height and thickness of this layer. The average height over the middle latitudes has been determined

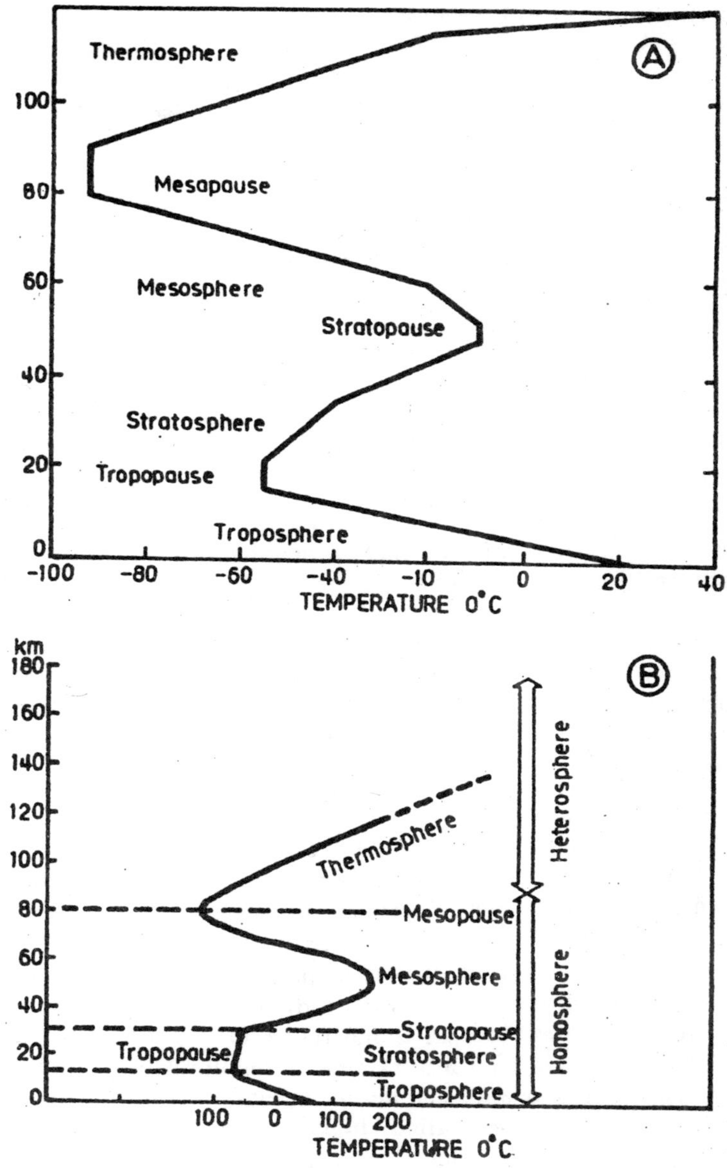

Fig. 1.1 : Stratification of the atmosphere. A- According to Bary and Charley. B- According to A.N. Strahler.

to be 25-30 km, whereas it is estimated to be 80 km by others. On an average the upper limit of the stratosphere is taken to be 50 km. There is also contrasting opinion about the change or no change of temperature with increasing height in this sphere. A few scientists believe that the stratosphere is isothermal *i.e.*, there is no change in temperature with increasing height while others hold that temperature gradually rises upward as it becomes 0°C or 32°F at the height of 50 km, the upper limit of the stratosphere which is known as *stratopause*. Though, the stratosphere is more or less devoid of major weather phenomena but there is circulation of feable winds and cirrus cloud in the lower stratosphere. The lower part of this layer is very important for life-forms in the biospheric ecosystem because there is concentration of ozone between the height of 15-30 km though ozone has been discovered upto the height of 80km.

The lower portion of the stratosphere having maximum concentration of ozone is called ozonosphere, which is confined between the height of 15 km to 35 km from sea level though the upper limit has been fixed at 55 km. Ozone (O_3) defined as 'a three-atom isotope of oxygen or merely a triatomic form of oxygen (O_3) is a faintly blue irritating gas with a characteristic pungent odour. The ozone gas is unstable because the creation and destruction of this gas is a gradual and continuous natural process.

It acts as a protective cover for the biological communities in the biosphere because it absorbs almost all of the ultra-violet rays of solar radiation and thus protects the earth's surface from becoming too hot. Recently, the researches have shown that there is gradual depletion of ozone gas in the atmosphere due to human activities. It may be pointed out that combining of atmospheric oxygen (O_2) with individual oxygen molecules results in the creation of ozone ($O_2 + O \rightarrow O_3$) whereas the breaking of ozone (O_3) into O_2 and O results in the depletion or destruction of ozone. The main culprits of ozone destruction are halogenated gases called chlorofluorocarbons, halons and nitrogen oxides. The chlorofluorocarbons, popularly known as CFCs, belong to the category of synthetic chemicals and are relatively simple compounds of the elements chlorine, fluorine and carbon and are initially stable compounds which do not have any toxic effect on life processes in the biosphere at ground level. These synthetic chemicals are widely used as propellants in spray can dispensers, as fluids in air conditioners and refrigerators etc. Chlorofluorocarbons, when used as propellants, are released into the air and are transported in the stratosphere by vertical atmospheric circulation. Chlorine when separated from chlorofluorocarbons reacts with water and thus depletes ozone rather breaks ozone into O_2 and O. Besides, nitrogen oxides released by supersonic jets which fly at the height of 18-22 km also depletes ozone. Depletion of ozone would result in the rise of temperature of the ground surface and lower atmosphere. This would cause global warming, acid rain, melting of continental glaciers and rise in sea level, skin cancer to white-skinned people, poisonous smogs, decrease in photosynthesis, ecological disaster and ecosystem instability.

(3) Mesosphere

Mesosphere extends between 50 km and 80 km. Temperature again decreases with increasing height. In fact, the rise of temperature with increasing height in the stratosphere stops at stratopause. At the uppermost limit of mesosphere (80 km) temperature becomes –80°C. This limit is called *mesopause* above which temperature increases with increasing height.

(4) Thermosphere

The part of the atmosphere beyond mesopause is known as thermosphere wherein temperature increases rapidly with increasing height. It is estimated that the temperature at its upper limit (height undecided) becomes 1700°C. It may be pointed out that this temperature cannot be measured by ordinary thermometer because the gases become very light due to extremely low density. That is why one does not feel warm when one stretches one's arm in the air. Thermosphere is divided into two layers viz.:

(i) Ionosphere, and

(ii) Exosphere.

(1) *Ionosphere* extends from 80 km to 640 km. There are a number of ionic layers (with increasing heights) in this sphere *e.g.* D layer, E layer, F layer, and G layer. *D layer* (between the height of 60 km–99 km) reflects the signals of low frequency radio waves but absorbs the signals of medium and high frequency waves. This layer disappears with the sunset because it is associated with solar radiation. E layer, also known as *Kennelly—Heaviside layer*, is confined in the height between 99 km–130 km. This layer reflects the medium and high frequency radio waves back to the earth. This layer is produced due to interaction of solar ultra-violet photons with nitrogen and nitrogen molecules and thus, it also disappears with the sunset. *Sporadic E layer* is associated with high velocity winds and is created under special circumstances. This layer reflects very high frequency radio waves.

E_2 *layer* is generally found at the height of 150 km and is produced due to reaction of ultra-violet solar photons with oxygen molecules and thus this layer also disappears during nights. F layer consists of two sub-layers *e.g.* F_1 and F_2 layers (150 km – 380 km) and are collectively called 'appleton layer'. These layers reflect medium and high frequency radio waves back to the earth. G layer (400 km and above) most probably persists day and night but is not detectable.

(2) *Exosphere* represents the uppermost layer of the atmosphere. In fact, we know very little about the atmosphere extending beyond 640 km height from the sea level. The density becomes extremely low and the atmosphere resembles a nebula because it is highly rarefied. The temperature becomes 5568°C at its outer limit but this temperature is entirely different from the air temperature of the earth's surface as it is never felt.

CHEMICAL COMPOSITION OF THE ATMOSPHERE

On the basis of chemical composition, the atmosphere is divided into two broad zones viz.:

(1) Homosphere,

(2) Heterosphere.

(1) *Homosphere* represents the lower portion of the atmosphere and extends upto the height of 90 km from sea level. The main constituent gases are oxygen (20.946%) and nitrogen (78.084%). Other gases are argan, carbon dioxide, neon, helium, krypton, xenon, hydrogen etc.

This zone is called *homosphere* because of the homogeneity of the proportion of various gases. In other words, the proportions of different gases are uniform at different levels in this zone. It may be pointed out that man is increasingly disturbing the natural proportions of gases through his ever increasing economic activities and modem technologies. For example, the proportion of carbon dioxide is rapidly increasing due to burning of fossil fuels (coal, petroleum and natural gas) and deforestation. The concentration of atmospheric carbon dioxide at the beginning of the industrial revolution (1860 A.D.) was fixed at 280 to 290 ppm (parts per million) by volume but now it has increased to 350-360 ppm (1988 A.D.)

Thus, registering an overall increase by 25 per cent from the pre-industrial level. On the other hand, the proportion of ozone gas is rapidly decreasing due to ever increasing production and consumption of CFCs (chlorofluorocarbons) and halogenated gases. On the basis of thermal conditions the homosphere has been divided into three layers viz. :

(i) Troposphere,

(ii) Stratosphere, and

(iii) Mesosphere.

(2) *Heterosphere* extends from 90 km to 10,000 km. Different layers of this sphere vary in their chemical and physical properties. There are four distinct layers of gases in this sphere:

(i) Molecular nitrogen layer is dominated by molecular nitrogen and extends upward upto the height of 200 km (90 to 200 km),

(ii) Atomic oxygen layer extends from 200 to 1100 km.

(iii) Further, upward there is helium layer which extends upto the height of 3500km.

(iv) Atomic hydrogen layer is the topmost layer of the atmosphere and extends upto the outermost limit of the atmosphere.

ELEMENTS OF WEATHER AND CLIMATE

Weather refers to the sum total of the atmospheric conditions in terms of temperature, pressure, wind, moisture, cloudiness, precipitation and visibility of a particular place at any given time. In fact, weather denotes short-term variations of atmospheric conditions and it is highly variable. On the other hand, climate is defined as aggregate weather conditions of any region in long-term perspective. According to Trewartha 'climate represents a composite of day to day weather conditions, and of the atmospheric elements, within a specified area over a long period of time.'

According to Critchfield 'climate is more than a statistical average; it is the aggregate of atmospheric conditions involving heat, moisture, and air movement. Extremes must always be considered in any climatic description in addition to means, trends, and probabilities.' According to Koeppen and De Long 'climate is a summary, a composite of weather conditions over a long period of time; truly portrayed, it includes details of variations-extremes, frequencies, sequences of the weather elements which occur from year to year, particularly in temperature and precipitation. Climate is the aggregate of the weather.' G.F. Taylor has maintained that 'climate is the integration of weather, and weather is the differentiation of climate. The distinction between weather and climate is, therefore, mainly one of time.' Temperature, pressure, wind, humidity, precipitation, cloudiness etc. are elements of weather and climate.

CONTROLS OF WEATHER AND CLIMATE

There are frequent changes in weather conditions. These changes from one day to the other or from one place to the other are due to variations in the quantity, intensity and distribution of the elements of weather and climate. Similarly, there is variation in climatic conditions from one area to the other area. The factors controlling the variations of the elements of weather and climate from one place to the other place and from one season to the other season are called controls of weather and climate. These factors include latitudes, altitudes, unequal distribution of land and water, ocean currents, air pressure and wind, mountain barrier, nature of ground surface, different types of atmospheric storms etc.

2

PRECIPITATION AND HUMIDITY

WATER-VAPOUR AN EVAPORATION

Humidity of the air refers to the content of water vapour present in the air at a particular time and place. On the other hand, water vapour is the gaseous form of water. Water vapour represents 2 per cent of the total composition of the atmosphere but this percentage varies both spatially and temporally as it ranges from zero to 5 per cent. Nearly 50 per cent of the total atmospheric vapour is concentrated in the lower atmosphere upto the height of 2000 metres. It may be mentioned that water occurs in three states viz. as solid (*e.g.*, ice, snow and frost), as liquid (*e.g.*, water), and in gaseous form (*e.g.* vapour). The presence of water vapour in the atmosphere is a vital factor for weather conditions of a particular region. The nature and amount of precipitation, the amount of loss of heat through radiation from the earth's surface, surface temperature, latent heat of the atmosphere, stability and instability of air masses etc. depend on the amount of water vapour present in the atmosphere. The atmospheric water vapour is derived through evaporation of water from oceans and seas, terrestrial lakes, land water bodies (tanks, ponds), rivers etc.

The process of transformation of liquid (water) into gaseous form is called evaporation. The amount and intensity of evaporation depend on aridity, temperature and velocity of winds. The higher the aridity, temperature and velocity of winds, the higher the rate and amount of evaporation because dry air with high temperature is capable of retaining more moisture (vapour) as dry air requires more time and moisture to become saturated. A stable air becomes saturated soon because there is no transfer of moisture while unstable air attains saturation quite late because there is much transfer of moisture.

There is more evaporation from the oceans than from the lands. There is maximum evaporation from the lands between 10°N and 10°S latitudes whereas maximum evaporation occurs from the oceans between 10°-200 latitudes in both the hemispheres.

The process of conversion of vapour into liquid (water) and solid form (ice, snow, frost) is called condensation. It is apparent that evaporation and condensation are opposite processes wherein the former involves conversion of liquid (water) into gaseous form (water vapour) while the latter refers to conversion of water vapour into liquid or solid form.

LATENT HEAT

Energy in the form of heat is required for the conversion of water into gaseous form (water vapour). Heat energy is generally measured in the unit of calorie, 79 calories are required to convert one gram of ice into water whereas 607 calories are needed for the conversion of one gram of water into water vapour. It is apparent that the potential energy of water is more than ice and that of vapour is more than water. This hidden amount of heat in water vapour is called *latent heat.*

In other words, the amount of heat spent during the process of evaporation is never lost rather it is always associated with water vapour. There are several evidences which demonstrate the use of heat energy at the time of evaporation. For example, (i) one feels cooling effect when one sits before fan or in shady open air after sweating in summer months because the sweats are evaporated, (ii) there is cooling effect when drops of sprit are kept on the palm of human hand because sprit is evaporated etc. On the other hand, heat energy is released at the time of condensation (conversion of vapour into liquid or solid form). This energy, released after condensation, is called *latent heat of condensation.*

HUMIDITY

As stated earlier, humidity refers to the content of water vapour present in the air in gaseous form at a particular time and place. The atmospheric humidity is obtained through various processes of evaporation from the land and water surfaces of the earth. The atmospheric humidity is of vital climatic significance because different forms of precipitation (dew, fog, rainfall, frost, snowfall, hailstorm etc.), atmospheric storms (cyclones) and turbulence etc. depend on humidity. The atmospheric humidity is expressed in a number of ways *e.g.* absolute humidity, specific humidity, relative humidity etc.

Humidity Capatity

The moisture content (humidity) of the air is measured in grain per cubic foot or in gram per cubic centimetre. Evaporation is the main mechanism through which water is converted into humidity (vapour). Since temperature and evaporation are directly positively related (evaporation increases with increasing temperature) and hence humidity and temperature are also directly positively related. The moisture retaining capacity or humidity capacity refers to the capacity of an air of certain volume at certain temperature to retain maximum amount of moisture content. Humidity capacity is directly positively related with temperature *i.e.*, higher the temperature, higher the humidity capacity and lower the temperature, lesser the humidity capacity. In other words, as air temperature increases, humidity also increases and *vice-versa*. For example, the humidity capacity of an air having the volume of one cubic foot and temperature of 30°F is 1.9 grain whereas it becomes 2.9 grain when temperature becomes 40°F (an increase of 10°F). It may be pointed out that the ratio of increase of humidity capacity also increases with increasing temperature (Table 2.1).

For example, the net increase of humidity capacity from 30°F to 40°F is one grain while it becomes 5 grain from 90°F to 100°F (Table 2.1). Similarly, humidity capacity becomes higher during summer months than during winter months and during daytime than nights.

The extent of land and water and wind velocity also influence humidity capacity. Oceanic and coastal areas record higher humidity capacity of air than the remote areas of the continents. Humidity capacity decreases from equator poleward. The air having moisture content equal to its humidity capacity is called *saturated air.*

Table 2.1 : Humidity Capacity

Temperature (0°F)	*Humidity Capacity (grain)*	*Difference at the interval of 10°F*	*Eastimated absolute humidity*
30	1.9	–	–
40	2.9	1.0	–
50	4.1	1.2	4.0
60	5.7	1.6	4.0
70	8.0	2.3	4.0
80	10.9	2.9	4.0
90	14.7	3.8	4.0
100	19.7	5.0	4.0

Specific Humidity

Specific humidity is defined as the mass of water vapour in grams contained in a kilogram of air and it represents the actual quantity of moisture present in a definite air. Specific humidity is seldom affected by changes in air pressure or air temperature because it is measured in the units of weight (grams). It is directly proportional to vapour pressure, which is 'the partial pressure exerted by water vapour in the air and is independent of other gases', and is inversely proportional to air pressure. Specific humidity decreases from equator poleward. For example, extremely cold and dry air over arctic region during winter generally has specific humidity of 0.2 gram per kilogram of air while it becomes as high as 18 grams per kilogram of extremely warm and moist air over equatorial regions. 'In a real sense, specific humidity is a geographer's yardstick of a basic natural resource-water-to be applied from equatorial to polar regions. It is a measure of the quantity of water that can be extracted from the atmosphere as precipitation' (A.N. Strahler and A. H. Strahler, 1978).

Relative Humidity

Relative humidity is defined as a ratio of the amount of water vapour actually present in the air having definite volume and temperature (*i.e.*, absolute humidity) to the maximum amount the air can hold (*i.e.*, humidity capacity). In other words, relative humidity is the proportion of absolute humidity of an air of definite volume at a given temperature to the humidity capacity of that air. Relative humidity is generally expressed as percentage. For example, if the humidity capacity and absolute humidity of an air having temperature of 20°C are 8 grains and 4 grains per cubic foot respectively.

There is inverse relationship between air temperature and relative humidity *i.e.*, relative humidity decreases with increasing temperature while it increases with decreasing temperature (Table 2.2).

Table 2.2 : Temperature, Humidity Capacity and Relative Humidity

Temperature (0°F)	*Humidity (grain per cubic foot)*	*Absolute humidity (grain per cubic foot)*	*Relative Humidity (%)*
30	1.9	1.9	100.0
40	2.9	1.9	65.5
50	4.1	1.9	46.3
60	5.7	1.9	33.3
70	8.0	1.9	23.7
80	10.9	1.9	17.4
90	14.7	1.9	12.9
100	19.7	1.9	9.6

When the humidity capacity and absolute humidity of the air are the same, the air is said to be saturated and the relative humidity becomes 100 per cent. Relative humidity changes in two ways viz., (i) if the absolute humidity increases due to additional evaporation or (ii) if the temperature of the air decreases so that humidity capacity also decreases. For example, the relative humidity of the air with 50°F temperature is 46%

because humidity capacity and absolute humidity are 4.1 grains and 1.9 grains respectively (Table 2.2) but if the temperature of that air decreases due to ascent to 40°P, the relative humidity becomes 65.5% because the humidity capacity decreases to 2.9 grains per cubic foot.

Importance of Relative Humidity : Relative humidity has a great climatic significance because the possibility of precipitation depends on it. High and low relative humidity is indicative of the possibility of wet (precipitation) and dry conditions respectively. The amount of evaporation also depends on relative humidity. Evaporation decreases with high relative humidity while it increases with low relative humidity. Relative humidity is directly related to human health and comfort. Very high (above 60%) and very low humidity is injurious to human health. This is why the equatorial regions with high temperature and high relative humidity and tropical hot deserts with very low relative humidity are unfavourable for human health. Relative humidity also affects the stability of different objects, buildings, electrical appliances, radio etc.

Distribution of Relative Humidity : The horizontal distribution of relative humidity on the earth's surface is zonal in character. Equatorial regions are characterized by highest relative humidity. It gradually decreases towards subtropical high pressure belts where it becomes minimum (between 25°-35° latitudes). It further increases poleward. The zones of high and low relative humidity shift northward and southward with northward and southward migration of the sun respectively. Seasonal distribution of relative humidity is largely controlled by latitudes. Maximum relative humidity is found during summer season between 30°N and 30°S latitudes while high latitudes record relative humidity more than average value during winters. There is maximum relative humidity in the morning whereas lowest value is recorded in the evening. Fig. 2.1 denotes zonal distribution of relative humidity.

The term which is applied to the presence of moisture in the atmosphere is *humidity*; this may be expressed either absolutely or relatively. The term *vapour concentration* is now preferred by the Meteorological Office for the former.

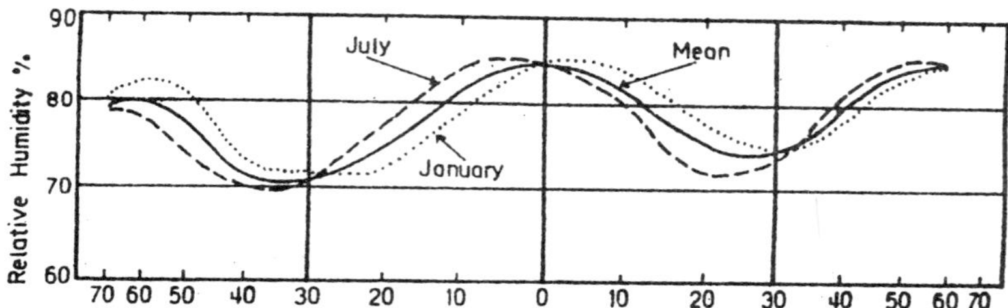

Fig. 2.1 : Zonal distribution of relative humidity.

Absolute Humidity

This is the actual amount of water-vapour present in a certain quantity of air, expressed in grammes per cu m or grains per cu ft. A mass of air, of a given temperature and pressure, can hold vapour up to a certain limited amount; when this limit is reached (that is, when as many molecules of water enter the air as are leaving it), the air is *saturated,* that is, the *saturation vapour pressure* has been reached. Saturated air at 10°C contains 9-41 grammes of vapour per cu m, at 20°C it contains 17-117 grammes, and at 30°C it contains 30-036 grammes. If air is not saturated, the mass of vapour required to bring it to the point of saturation is called the *saturation vapour-pressure deficit.*

The absolute humidity over land areas tends to be highest near the Equator throughout the year, and to be lowest in the great high-pressure systems covering central Asia in winter and in the Antarctic. Elsewhere it is extremely variable. Paradoxically, it is often extremely high in the hot deserts; much of the trying quality of Aden or the Persian Gulf in August is die result of the 'sticky-heat', the combination of high humidity and high temperature.

Vapour-pressure : Water-vapour exerts a definite pressure. The maximum vapour-pressure at any temperature occurs when the air is saturated; tables are available giving the vapour-pressure of saturated air at different temperatures. The average vapour-pressure in England on a summer afternoon amounts to about 15 millibars (approximately equivalent to 11.4 mm, 0.45 in of mercury), at the Equator about 30 millibars. Absolute humidity can, therefore, be expressed either in terms of the mass of water-vapour per cubic unit or in terms of the pressure it exerts.

Other practical expressions of moisture content are *specific humidity* (the ratio of the weight of water-vapour present to the total weight of the air with its moisture content, usually expressed as grammes per kilogram) and the *mixing ratio* (the ratio of the weight of water-vapour present to the weight of the air less its moisture content). There is no change in absolute humidity unless vapour is actually added to or removed from a body of air. If the amount of vapour present is expressed as a percentage of the total amount that would be present were the air saturated, the relative humidity is obtained. Saturated air at 20°C contains 17·117 grammes of vapour per cu m; if, for example, a mass of air at 20°C contained only 8·262 grammes, the relative humidity would be (8·262 × 100) ÷ 17·117, approximately 48 per cent.

Relative humidity varies not only with the absolute humidity, but with the temperature of the air, for as the temperature rises so the relative humidity falls. If a mass of air, saturated at 4°C (that is, with relative humidity = 100 per cent), is warmed, at 10°C the relative humidity falls at 71 per cent, at 15°C to 51 per cent, and at 32°C to 19 per cent. Conversely, when a mass of unsaturated air is cooled, the relative humidity rises until it reaches the point of saturation at 100 per cent. Beyond this, any further cooling normally causes condensation of the excess vapour which the air is no longer able to hold, in the form of minute drops of water. This critical temperature is known as the *dew-point*.

Measurement of Relative Humidity

The most commonly used instrument for determining relative humidity is a 'wet and dry-bulb thermometer'. This consists of a pair of mercury thermometers, mounted side by side on a stand; round one of the bulbs is tied a cotton or muslin bag, kept moist by a wick leading down into water. If the air is saturated, both the thermometers show the same reading. If the air is not saturated, evaporation will take place from the moist cloth and so cool the wet-bulb, since latent heat (that is, the heat expended or consumed in doing work by changing the physical state of a body) is used up in the process of vaporization. The difference in the temperatures recorded by the two thermometers is noted, and the relative humidity is obtained from tables which are available. Thus, if the dry-bulb thermometer reads 21°C and the wet-bulb 15°C, the relative humidity can be seen from tables to be 54 per cent.

To ensure that maximum evaporation is taking place from the wet-bulb, various refinements have been devised. One method is the mounting of the thermometers in a sling or in a kind of rattle, so that they can be swung around; these are *whirling psychrometers*. The *Assman psychrometer* incorporates a small electric fan.

Continuous records of relative humidity, although not particularly accurate, are made by a *hygrograph*. This consists of strands of human hair, which lengthen and shorten according to the humidity; these minute changes are amplified and transferred to an inked pen, which traces a record upon a chart fixed to a rotating drum.

THE HUMIDITY OF AIR-MASSES

An important effect of the presence of water-vapour in a body of air is the reduction of its density. A quantity of water-vapour is lighter than the same volume of dry air, in the proportion of about 5 to 8. When dry air takes up water-vapour by means of evaporation, this is not just a net addition, but it replaces an equivalent volume of air. The important result follows that moist air is less dense, that is, lighter, than an equal volume of dry air at the same temperature and pressure.

One other physical principle must be stressed at this point. If air is compressed, not only does its density change, but its temperature also; if a bicycle tyre is blown up, the valve becomes uncomfortably warm because compression causes dynamic heating. Expansion, on the other hand, results in dynamic cooling. The usual way in which an air-mass can expand is when it ascends bodily into the upper atmosphere, since there will be a smaller amount of air above it and therefore lower pressure. When an airmass undergoes a change of temperature without any heat being lost or gained by the air-mass from outside, it experiences *adiabatic warming* (on compression) or *adiabatic cooling* (on expansion). This is in contrast to diabatic change, when a temperature change at the earth's surface involves the mixing of air, with a definite gain or loss of heat from its surroundings.

Lapse-rates

The average decrease of temperature in still air with altitude (the *environmental lapse-rate* or vertical temperature *gradient*) is, though extremely variable, about 0·6°C per 100 metres (3·3°F per thousand feet). When an unsaturated air-mass ascends vertically it expands, and therefore cools adiabatically. This loss of temperature is known as the *dry adiabatic lapse-rate*, a meteorological constant, which is a temperature decrease of 1°C for every 100 m of ascent (5.4°F for every thousand feet). If a saturated air-mass ascends vertically, it also expands and cools, and as it is saturated to start with, some of its water-vapour is immediately condensed. This means that a certain amount of latent heat is liberated, which reduces the rate at which the ascending airmass cools. This *saturated* (or *wet*) *adiabatic lapse-rate* can lie anywhere between 0·3°C and 0.9°C per 100 m of ascent, according to the amount of water vapour present and to the temperature of the air.

Table 2.3

Metres	*(In degrees Centigrade)*		
1200	12.8	9.0	15.2
900	14.6	11.0	16.4
600	16.4	14.0	17.6
300	18.2	17.0	18.8
Sea-level	20.2	20.0	20.0
	Environmental atmospheric conditions in still air (average 0.6°C per 100 m)	Column of rising unsaturated air (1°C per 100m)	Column of rising saturated air (at 0.4°C per 100m)

Thus, an airmass of about 30°C may contain so much water-vapour, and therefore release so much latent heat, that the saturated adiabatic lapse-rate is only 0-4°G per 100 m, while in a very cold air-mass or at high altitudes there may be so little water-vapour that the saturated adiabatic lapse-rate will not differ materially from the dry adiabatic lapse-rate. The following table gives conditions for various lapse-rates.

Instability

'Pockets' of air are forced to rise in several ways. One is when local heating of the earth's surface takes place; the overlying air-mass is warmed by conduction and a vertical convection current is set up. Another cause is when a wind blows against a mountain-side, mechanically forcing the 'pocket' up the slope. A third involves the rising of one mass of air above another at a frontal surface. In most cases of rising air a vortical element is introduced, a form of convergent rotation.

When a 'pocket' of air is warmer, and therefore lighter, than its surroundings, it will rise bodily; such a mass is said to be *absolutely unstable*, or in *unstable equilibrium*. Instability is much more common in the case of saturated than of unsaturated air, since it cools much less rapidly, and so remains warmer than its surroundings. A warm, damp air-mass may therefore rise to great heights, and cause very unstable atmospheric conditions, building up great clouds, and possibly causing heavy rainfall, hail and thunderstorms (Fig. 2.2). When ultimately the air-mass reaches a height at which it has the same temperature as the surrounding air, vertical ascent ceases; it is now in *neutral* (or *indifferent*) *equilibrium*.

If a mass of dry, or unsaturated, air rises as a wind up a mountain slope, it cools at the dry adiabatic lapse-rate until it reaches its dew-point. Then condensation begins, and it now cools at the saturated adiabatic lapse-rate. This may mean that even though the initial cause of the uplift was mechanical, the lower rate of cooling results in the air-mass remaining relatively warmer than its surroundings, and so it will continue ascending of its own accord. This state of affairs, known as *conditional instability*, is a common cause of storms and heavy rain; the term is derived from the fact that instability is conditional upon the presence of sufficient water-vapour. It is not necessary to have a forced orographic ascent for conditional instability to develop, although this is the commonest cause; it may develop whenever the general environmental lapse-rate lies between the dry and saturated adiabatic rates.

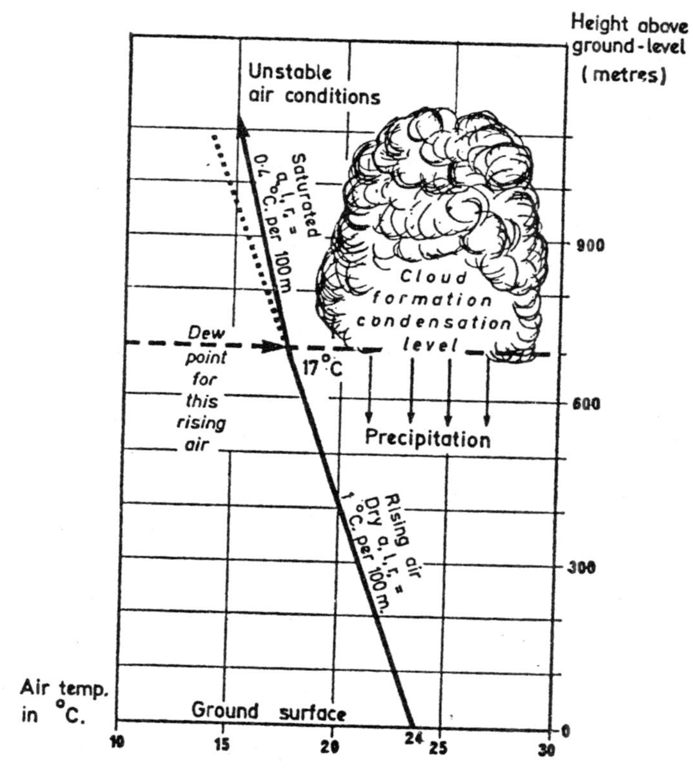

Fig. 2.2 : The formation of cloud and precipitation in an ascending air-mass.

If, however, the mass of dry air which is rising in the form of a wind has a lapse-rate greater than that of the surrounding air, and does not reach dew-point, it will in due course become cooler than its surroundings. As soon as the mechanical force ceases (that is, when the wind drops) it tends to sink back to lower levels, since it is heavier, that is, denser, than its surroundings. This is known as *stable equilibrium* or simply as *stability*. A mass of air, which becomes conditionally unstable when forced to rise over a relief barrier or over a cold air-mass at a front, is in a state of *potential instability*.

Stability and instability have been discussed in some detail, since the vertical movement of air-masses of differing temperature and humidity is intimately associated with atmospheric disturbances. This is, moreover, the most potent cause of precipitation.

CONDENSATION AND ASSOCIATED FORMS

The transformation of gaseous form of water (*i.e.,* water vapour) into solid form (ice) and liquid form (water) is called condensation. In other words, the process of change of water vapour into liquid form is called condensation. It is evident that condensation is opposite to evaporation. The process and mechanism of condensation depends on the amount of relative humidity present in the air. The air having MX) per cent relative humidity is called saturated air. An air may become saturated in two ways *e.g.* either (i) the absolute humidity at a given temperature is raised to equal the humidity retaining capacity of the air or (ii) the temperature of the air is reduced to such an extent that the humidity capacity becomes equal to its absolute humidity. For example, the humidity capacity and absolute humidity of an air with 60°F temperature are 5.7 grains and 4.1 grains per cubic foot respectively (relative humidity being 72 per cent), if the air is cooled to 50°F temperature, the humidity capacity decreases to 4.1 grains which is equal to absolute humidity of 4.1 grains per cubic foot, thus the air becomes saturated as relative humidity becomes 100 per cent, and hence condensation begins. The temperature at which an air becomes saturated is called dew point. It may be pointed out that condensation will begin only when the air is supresaturated *i.e.,* if the relative humidity exceeds 100 per cent, and this can be achieved only when the air is further cooled. If dew point is above freezing point (32°F), condensation will occur in liquid form (*e.g.*, dew, fog, rainfall etc.) but if dew point is below freezing point, condensation occurs in solid form (*e.g.*, frost, ice, snow, hailstorm etc.).

It is apparent that condensation depends on (i) the percentage of relative humidity of the air and (ii) the degree of cooling of the air. The air becomes cool when it rises while it gets heated when it descends. Thus, the ascending air may bring moist weather while descending air causes dry condition. Much cooling of the air is required in hot arid regions before dew point is reached. On the other hand, very little cooling causes condensation in humid regions. The heat released at the time of condensation is called latent heat of condensation.

RAINFALL

Origin of Rainfall

The presence of warm, moist and unstable air and sufficient number of hygroscopic nuclei are prerequisite conditions for rainfall. The warm and moist air after being lifted upward becomes saturated and clouds are formed after condensation of water vapour around hygroscopic nuclei (salt and dust particles) but still there may not be rainfall unless the air is supersaturated. The process of condensation begins only when the relative humidity of ascending air becomes 100 per cent and air is further cooled through dry adiabatic lapse rate but first condensation occurs around larger hygroscopic nuclei only. Such droplets are called cloud droplets. The aggregation of large number of cloud droplets forms clouds. These cloud droplets are so microscopic in size that they remain suspended in the air. Rainfall does not occur unless these cloud droplets become so large due to coalescence that the air becomes unable to hold them. This is why, some times the sky is overcast by thick clouds but there is no rainfall. If by chance these cloud droplets fall downward they are evaporated before they reach the ground surface. Rainfall occurs only when cloud droplets change to raindrops. There are two possible processes of change of cloud droplets into raindrops.

(1) If warm and moist air ascends to such a height that condensation begins below freezing point, then both, water droplets and ice droplets, are formed. The water droplets are evaporated because of difference of vapour pressure between them and ice droplets and there is

condensation of evaporated water around ice crystals which go on increasing in size. If they become sufficiently large in size, they cannot be held in suspension by the air and consequently they begin to fall down. If the temperature above the ground is high they fall in the form of raindrops.

(2) The suspended cloud droplets in the clouds are of different sizes. These cloud droplets collide among themselves at varying rates due to difference in their sizes and thus form large droplets. In the process several cloud droplets are coalesced to form raindrops. When they become so large in size that ascending air becomes unable to hold them, they fall down as rainfall.

The diameter of a raindrop is upto 5mm and one raindrop contains about 8,000,000 cloud droplets. Rain drops fall down at the velocity 200 times greater than cloud droplets. When raindrops become very large and fall down at greater speed (more than 30 kilometres per hour), they are split in the transit but give heavy downpour. When the air ascends slowly, the process of condensation is also very slow and hence small raindrops are formed and the resultant rainfall is drizzle but if the air ascends hurriedly with greater speed. Very large raindrops are formed and resultant rainfall is heavy downpour. When condensation occurs below freezing point, the resultant precipitation is in solid form and is called snowfall.

Theories of Rainfall

Various theories of precipitation and rainfall have been put forth from time to time but the riddle of raindrop formation still remains unresolved. Very generalized two processes and mechanisms of raindrop formation have been outlined above. The early theories related to the formation of raindrops and their growth may be briefly summarized in the following manner:

(i) The raindrops are differently electrically charged and thus they are coalesced by electrical attraction and grow in large size. This theory is opposed on the ground that the distances between raindrops are so large and difference between electrical charges is so small that coalescence of raindrops due to this mechanism is not possible. In other words, the coalescence and growth of raindrops due to differential electrical charges are not possible.

(ii) The large rain drops capture small rain drops and thus become further large in size but the observations have revealed that there is regular pattern in the size and distribution of drops in the clouds. In other words, generally most of the drops are more or less uniform in size as their diameters range between 20 to 30 micrometres and only a few drops are larger than 80 micrometres.

(iii) There is variation in saturation vapour pressure with varying temperature. In such condition the atmospheric turbulence brings warm and cold cloud droplets in close conjunction, with the result there is supersaturation of air with reference to the surface of the cold cloud droplets and undersaturation of air with reference to the surface of warm cloud droplets and growth of cold droplets at the expense of warm droplets. This situation causes evaporation of warm droplets. This theory is opposed on the ground that the difference of temperature of cloud droplets is not so great that this differential mechanism may operate.

(iv) Raindrops grow around very large condensation (hygroscopic) nuclei but it is argued that no doubt the process of condensation around exceptionally large hygroscopic nuclei is very rapid but their further growth cannot be explained on this mere ground of size of cloud droplets.

The theories of precipitation and rainfall fall in two main categories *e.g.* (1) rapid growth of raindrops due to growth of ice crystals at the cost of water droplets and (2) rapid growth of raindrops due to coalescence of small water droplets by the sweeping actions of falling drops.

(1) *Cloud Instability Theory of Bergeron Findeisen :* For Bergeron, an eminent Norwagian meteorologist, postulated his theory of precipitation, known as 'cloud

instability theory' or 'ice crystal theory' in 1933. The core of the theory is based on the concept of mechanism of the growth of raindrops. According to him water droplets and ice crystals are found together in unstable clouds when temperature is below freezing point. The theory is based on commonly accepted fact that "the relative humidity of air is greater with respect to an ice surface than with respect to water surface". With the fall of air temperature below 0°C the atmospheric vapour pressure decreases more rapidly over ice surface than over water surface with the result saturation vapour pressure becomes greater over water surface than ice surface when the air temperature ranges between –5°C and –25°C and the difference between saturation vapour pressure over water and ice surfaces exceeds 0.2 mb. In such condition, when air temperature ranges between –5°C and –25°C, water droplets become super-saturated.

If ice crystals and supercooled water droplets exist together in a cloud, then the water droplets are evaporated and resultant vapour is deposited on to the ice crystals.

It may be pointed out that the formation of ice particles requires freezing nuclei (*e.g.* fine soil particles, meteoric dust etc.) in the same manner as the formation of water droplets requires the presence of hygroscopic nuclei. Slowly and slowly ice crystals grow in size as the deposition of vapour derived through evaporation of supercooled water droplets on their surfaces continues. Ice crystals then aggregate due to their mutual collision and thus they form large snow Hakes.

The aggregation of ice crystals is more prevalent when air temperature ranges between 0°C and –5°C. When the ice crystals become large (snow flakes) and their falling velocity exceeds the velocity of rising air currents, they fall downward. When the falling ice crystals pass through a thick layer of air with temperature more than 0°C, they are changed into raindrops and thus begins rainfall.

(2) *Collision Theory* : Though the Bergeron process of the origin of precipitation and rainfall satisfied most of the observed facts but it could not explain the mechanism of rainfall in the tropical areas where cumulus clouds over the oceans give copious rains when they are only 2000m thick and the air temperature at their top is 5°C or even more. It is thus evident that ice crystals do not help in the formation and growth of raindrops of large size in warm clouds.

Thus, collision theory involving collision, coalescence and sweeping for the formation and growth of raindrops was postulated by several meteorologists. According to some meteorologists atmospheric turbulence causes collision of cloud droplets. Due to collision they coalesce and grow in size. This concept suffers from two shortcomings *e.g.*:

(i) Collision, may cause splitting and scattering of cloud droplets rather than their aggregation due to coalescence, and

(ii) There is little and often no precipitation from highly turbulent clouds. Longmuir suggested modifications in the 'general coalescence theory' in order to plug its loopholes as mentioned above. According to him the terminal velocities of falling drops are directly related to their diameters.

In other words, larger drops fall with greater velocity than smaller drops. Thus, large drops absorb smaller droplets. Smaller droplets are also swept by larger droplets. All these lead to increase in the size of larger droplets which become raindrops which fall as rains because they cannot be held in suspension by rising air currents.

Rain is the most common form of precipitation. For rainfall, it is necessary that moist air must ascend, saturate (relative humidity 100 per cent) and condense. Adiabatic cooling due to upward movement of air is by far the most important mechanism of condensation and related precipitation including rainfall. It is apparent that upward movement is a prerequisite condition for cloud formation and rainfall. Thus, precipitation and rainfall are classified on the basis of conditions and mechanisms of upward movement of air.

There are three ways in which air is forced to move upward and thus cools according to adiabatic lapse rate *e.g.* (1) due to heating of ground surface the air being heated expands and rises upward in the form of convection currents, the mechanism is known as thermal convection, (2) ascent of air over an orographic barrier, and (3) uplift of air associated with low pressure system, known as cyclonic or frontal ascent.

It may be pointed out that it is not necessary that all these three factors work independently in relation to the ascent of air. Some times, more than one factor are operative. In such situation, the form of precipitation is determined on the basis of dominant factor. Thus, precipitation and rainfall are classified into the following three types:

(1) convectional rainfall, occurring due to thermal convection currents caused due to insolational heating of ground surface.

(2) orographic rainfall, occurring due to ascent of air forced by mountain barrier, and

(3) cyclonic or frontal rainfall, occurring due to upward movement of air caused by convergence of extensive air masses.

TYPES OF RAINFALL

When minute droplets of water are condensed from water-vapour in the atmosphere on to nuclei, they may float as clouds, If, however, the droplets coalesce, they form larger drops, which, when heavy enough to overcome by gravity ascending air currents within a cloud, fall as rain. One theory maintains that the presence of ice spicules acting as nuclei is necessary for precipitation to occur.

Interesting experiments to stimulate rainfall have been made during recent years, particularly in the U.S.A. Their basic principle is 'seeding', that is, the introduction into cumulus clouds of fine particles, usually of solid carbon dioxide ('dry ice'), silver iodide, or even volcanic dust, from aircraft or balloons in order to provide nuclei for the coalescence of droplets, so forming raindrops. Some success is claimed for these methods in the Midwestern states of America, which are particularly liable to disastrous droughts.

For condensation and precipitation to occur naturally, the appreciable ascent of an air-mass is essential. This ascent is brought about in three main ways, hence there are three main types of rainfall:

(i) *Convectional* rainfall due to surface heating,

(ii) Orographic or relief rainfall due to a forced ascent over land, particularly over a high range of hills and

(iii) *Frontal* or *cyclonic* rainfall, when either a mass of warm air overruns cold air or the latter undercuts the former.

(i) *Convectional rainfall :* This most commonly results from the updraught of air (a 'thermal'), which, having been warmed by conduction from a heated land surface, expands and rises, and in so doing is adiabatically cooled. The local heating starts the whole process and is therefore known as a 'trigger effect', but once the uprush of air has started, possibly in the form of a circular or elliptical vortex ring the resultant conditional instability will allow it to carry on, even when heating is cut off, as when the developing cloud drifts across the sun and there is a sudden feeling of chill before the storm breaks. The vortex ring will expand, partly because of the decrease of air pressure with altitude, partly because the ambient air is 'sucked in' and up through it. Where the surface air is particularly warm and humid and the upper air is abnormally cold, a condition of extreme instability results, with powerful turbulent vortical up-currents. Cumulonimbus clouds with an immense vertical range then form, in which icing conditions will develop, and from which heavy rain may fall. This is especially marked in the 'hot tower' of a tropical low pressure system.

Convectional rainfall occurs throughout the year near the Equator, where constant high temperatures and humidity produce this type of rainfall almost daily in the afternoon. With increasing distance from the Equator the rainfall is associated more markedly with summer, and both the total amount and the duration of the rainy season decrease as the hot deserts are approached.

In middle latitudes convectional rainfall occurs in early summer, when the upper atmosphere is still cool following winter, but when heating of the earth's surface is becoming active. The rainfall in the continental interiors in early summer is mainly of this type.

Thunderstorms : A development of convectional overturning under extreme conditions of instability may produce thunderstorms. As the towering cumulonimbus cloud advances across the sky, the barometer falls markedly, and a strong fresh wind relieves the sultriness which usually precedes a storm. Heavy rain or hail falls as the cloud passes overhead, accompanied by lightning, either 'fork' or the common 'sheet' lightning of summer.

When condensation takes place at the level of saturation, the uprush of air is still so great that the droplets are carried upward; cases have even been known whereby pilots of aircraft baling out by parachute have been carried upward for some distance by these air currents. The largest possible droplet that can form is 5'5 mm in diameter, for beyond that size it becomes unstable and breaks into one or more droplets; this may go on repeatedly, but if the rising air currents decrease in strength, the drops will fall to earth.

One theory assumes that each time a drop breaks up, the resulting droplets assume a positive charge of static electricity, the surrounding air taking a negative charge. The upper part of the cloud becomes positively charged, the lower part and the surrounding air negatively charged, -with a further small subsidiary positive charge near the base of the cloud. A difference of as much as 100 million volts may develop, and a lightning flash represents the re-uniting of the separate charges along a direct channel through the atmosphere, either within the cloud, or from cloud to ground. The thunder which follows the discharge is explained in terms of the disturbance of the air particles in the atmosphere, producing a noise reflected to earth from the cloud surfaces. But this explanation of thunderstorms is much too simple; scientists who have spent many years studying these features are still not certain of what really happens. Much work has been done particularly by B. J. Mason; two important concepts have been suggested by him. The first involves the creation of vertical 'cells' within the thunder cloud, extending throughout its full height, with a very strong vortical updraught and the formation of ice crystals in the upper part, and a corresponding downdraught, stimulated by frictional drag. The other is that when ice particles of initially different temperatures collide, the 'warmer' piece acquires a positive charge, the 'colder' piece an equal negative charge. A collision between super-cooled water droplets and hailstones, both present in a cell, will also cause a separation of positive and negative charges.

Other types of thunderstorms occur in addition to the convectional or thermal varieties, though these are the most common. *Frontal thunderstorms* in middle latitudes are associated with the movement of a cold front, particularly with an accompanying line-squall. Occasionally, they are associated with a warm front when the warm air-stream is particularly unstable. Thunderstorms frequently accompany orographic rainfall, particularly in the Tropics where markedly moist and warm air-masses rise sharply over steep mountain ranges, as in Java in Indonesia.

The close association between the relief and the mean annual rainfall totals will be noted. The solid area indicates the relief profile. The distance from Workington to Stockton is 145 km (90 miles).

(ii) *Orographic rainfall :* This type of rainfall occurs when air is forced to ascend the side of a mountain range. This may 'trigger off' conditional instability; it may cause convergence and uplift; it may increase precipitation by retarding the rate at which a depression moves; and it may steepen a cold front by friction. It is found wherever hills lie parallel to the coast over which blow moist winds from the sea. There is a pronounced difference between the windward and leeward sides of the mountains; the markedly drier leeward side is the rain-shadow (Fig. 2.3). The orographic factor usually increases precipitation produced by other causes. Thus, the occurrence of high relief along the margins of western Britain causes an intensification of frontal (depressional) precipitation.

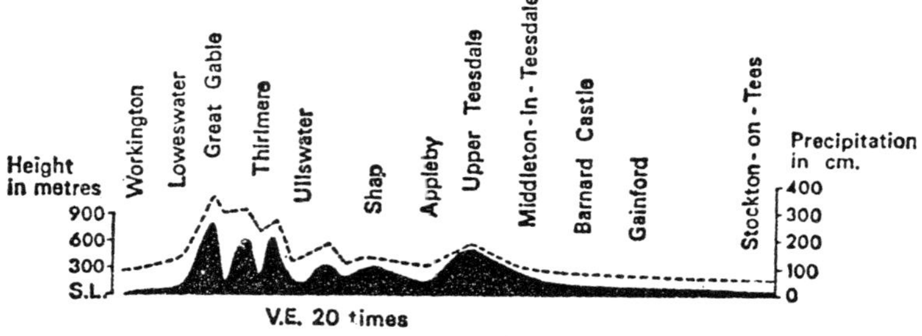

Fig. 2.3 : Relief and rainfall profile across northern England.

(iii) *Frontal rainfall :* This type of rainfall occurs along the frontal zones of convergence, notably at the Intertropical Convergence Zone and at the Polar Fronts. In the case of the ITCZ the diurnal convectional rhythm is strongly marked, superimposed upon frontal influences. Patterns of air convergence in the ITCZ waves may, under the prevailing warm, humid conditions, cause the build-up of massive cumulonimbus clouds, with torrential rain and thunderstorms.

In the case of fronts associated with local low-pressure systems which usually travel from west to east in middle and high latitudes, continuous drizzling rain falls in a broad belt along the warm front and in the warm sector, while more concentrated squally showers fall as the cold front passes. The rainfall is intensified by the effect of relief as the depression crosses the coasts, as in western Britain, Norway and British Columbia. Occasionally small depressions of great intensity, with a warm exceptionally humid airstream in the warm sector, may produce torrential downpours rare in middle latitudes; 28

cm (11 in) of rain fell at Martinstown near Dorchester (on 18.19 July 1955) during a period of only 9.5 hours. The actual days of heaviest rain recorded in England were 26-67 cm (9.56 in) at Bruton in Somerset (28 June 1917), 23.88 cm (9.40 in) at Cannington in Somerset (18 August 1924), and 23.11 cm (9.1 in) at Simonsbath in Somerset (15 August 1952).

Aridity

The causes and distribution of aridity are in effect complementary to those of rainfall. Although in a sense it is a negative feature, aridity (both seasonal and annual) is a major climatic fact, and its importance is seen in the delimitation of climatic types in the following chapter.

Aridity occurs on leeward slopes, that is, in rain-shadow areas. It is also found where an area is subject to dry land-winds, or to winds blowing from cooler to warmer latitudes, which exercise a drying effect. Again, it occurs where stationary anticyclones appear either on a small scale in middle latitudes or on a large scale in the subtropical land areas and mid-latitude continental interiors, the great air-mass source-regions. Aridity also occurs where the atmosphere is at a constantly low temperature and can contain little vapour, as in the Tundra and the polar regions.

It is important to remember that aridity depends not merely on the amount of precipitation, but on its effectiveness, hence the term *precipitation-effectiveness* (PE). Various empirical formulae, involving both precipitation and temperature, have been devised to obtain values, especially for use in climatic classification.

(1) Covectional Rainfall

The principal motivating force behind the ascent of warm and moist air is thermal convection caused by heating of the ground surface through insolation. Two conditions are necessary to cause convectional precipitation and rainfall *e.g.* (i) abundant supply of moisture through evaporation to the air so that relative humidity becomes high, and (ii) intense heating of ground surface through incoming shortwave electromagnetic solar radiation (say, insolation heating). The mechanism of convectional rainfall may be explained in the following manner.

The ground surface is intensely heated due to enormous amount of heat received through solar radiation during daytime, with the result the air coming in contact with warm ground surface also gets heated, becomes warm, expands, and ultimately rises upward. The ascending warm and moist air cools according to dry adiabatic lapse rate (decrease of temperature at the rate of 10°C per 1000 metres). The cooling of ascending air increases its relative humidity. The moist air becomes saturated soon (relative humidity becomes 100 per cent) and further ascent of air beyond saturation level causes condensation and cloud formation (cumulo-nimbus clouds) and thus rainfall starts. The air still continues to rise and in the process further cools but at moist adiabatic lapse rate or retarded adiabatic rate (decrease of temperature at the rate of 5°C per 1000 metres) due to addition of latent heat of condensation to the ascending air released after condensation of atmospheric vapour. When the ascending air reaches such height where its temperature matches with the temperature of surrounding air, the process of condensation is more activated and hence cumulo-nimbus clouds are formed and there begins instantaneous heavy rainfall (Fig. 2.4).

Since the ascending moist air (convectively motivated) cools soon after rising to very little height, causing immediate saturation and condensation, the convectional rainfall occurs in the form of heavy downpour. It is also apparent from the above description that convectional rainfall is a warm weather phenomenon and is associated with lightning and cloud thunder. Convectional rainfall mainly occurs in equatorial regions of low latitudes where daily heating of ground surface upto noon causes convection currents. Consequently, the sky becomes overcast by 2-3 P.M. daily causing pitch darkness and heavy rains and the sky becomes clear by 4 P.M. Thus, the convectional rainfall in the equatorial region is a daily regular feature.

Convectional rainfall also occurs in the tropical, subtropical and temperate regions in summer months and in the warmer parts of the day. The following are characteristic features of convectional rainfall:

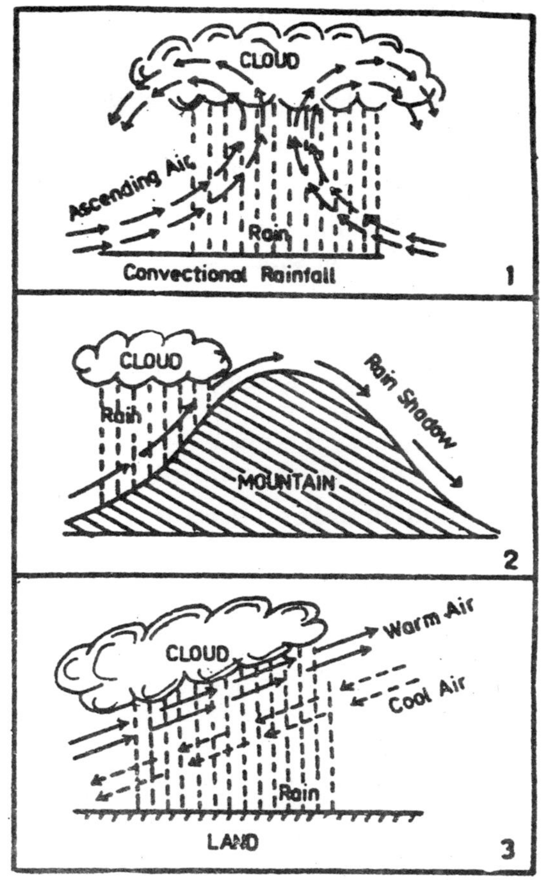

Fig. 2.4 : Types of rainfall : (1) convectional rainfall. (2) orographic rainfall, and (3) cyclonic or frontal rainfall.

(i) It occurs daily in the afternoon in the equatorial regions.

(ii) It is of very short duration but occurs in the form of heavy showers (heavy downpour).

(iii) it occurs through thick dark and extensive cumulonimbus clouds.

(iv) It is accompanied by cloud thunder and lightning.

(v) Though much of the rainfall becomes runoff and is drained off in the form of overland flow to the streams but still there is sufficient moisture in the soils due to daily rainfall in the equatorial regions. Out side equatorial regions convective rainfall is of little significance to crop growth because most of rainwater is drained to the streams through surface runoff which causes severe rill and gully erosion resulting into enormous loss of loose soils.

(vi) Convective rainfall supports luxurious evergreen rainforests in the equatorial regions.

(vii) Convective rainfall in the temperate regions is not in the form of heavy showers rather it is slow and of longer duration so that most of rainwater infiltrates into soils. Here rains are always in summers.

(viii) Convective rainfall in hot deserts is not regular but is irregular and sudden.

(2) Orographic Rainfall

Orographic rainfall occurs due to ascent of air forced by mountain barriers. The mountain barriers lying across the direction of air flow force the moisture laden air to rise along the mountain slope and thus lifted air mass cools according to dry adiabatic lapse rate (decrease of temperature at the rate of 10°C per 1000 metres) which increases the relative humidity of the air. The ascending air becomes saturated after reaching certain height and condensation begins around hygroscopic nuclei. The addition of latent heat of condensation to the air causes it to move further upward and cool at moist adiabatic lapse rate (decrease of temperature with increasing height at the rate of 5°C per 1000 metres). Thus, ascending air continues to yield precipitation with increasing height. It is apparent that mountain barriers produce trigger effect for the moist air to ascend, cool and become unstable. The slope of the mountain facing the wind is called windward slope or onward slope and receives maximum precipitation while the opposite slope is called leeward slope or rain shadow region because the ascending air after crossing over the mountain barrier descends along the leeward slope and thus is warmed at dry adiabatic lapse rate (increase in temperature with decreasing height at the rate of 10°C per 1000 metres). Consequently, the humidity capacity of the descending air increases resulting into substantial decrease in relative humidity. Secondly, the moisture present in the air is already precipitated on the windward slope and thus there is very little precipitation on the leeward slope. Most of the world precipitation occurs through orographic rainfall.

The following conditions are necessary for the occurrence of orographic rainfall.

(i) There should be mountain barrier across the wind direction, so that the moist air is forced on obstruction to move upward. If the mountain barriers are parallel to the wind direction, the air is not obstructed and no rainfall occurs. For example, Aravallis ranges running in southwest-northeast direction are parallel to the Arabian Sea Branch of south-west Indian monsoon and hence Rajasthan receives very low amount of rainfall.

(ii) If the mountains are very close and parallel to the sea coasts, they become effective barriers because the moisture laden winds coming from over the oceans are obstructed and forced to ascend and soon become saturated. For example. Coast Range mountains situated on the western margins of North America are parallel to the Pacific coast. Similarly, the situation of the Western Ghats in India presents ideal conditions for orographic rainfall.

(iii) The height of mountains also affects the form and amount of orographic rainfall. If the mountains are very close to the sea coast, even low height can be effective barrier and can yield sufficient rainfall because the moist air becomes saturated at very low height. On the other hand, the inland mountains should be of higher height because the air after covering long distances loses much of its moisture content.

(iv) There should be sufficient amount of moisture content in the air.

The following are the characteristic features of orographic rainfall:

(a) The windward slope, also called as rain slope, receives maximum amount of rainfall whereas leeward side of the mountain gets very low rainfall. For example, Mangalore located on the western slope (windward slope) of the Western Ghats receives mean annual rainfall of above 2000 mm whereas Bangalore situated in the rain shadow region gets only 500 mm of mean annual rainfall. The southern slopes of the Himalayas receive mean annual rainfall of more than 2000 mm whereas the northern slope receives only 50 mm of mean annual rainfall. Similarly, the western slopes of the Coast Ranges of North America receive more than 2000mm of mean annual rainfall while the eastern slopes fall in rainshadow region.

(b) There is maximum rainfall near the mountain slopes and it decreases away from the foothills. For example, the cities and towns located at the southern slopes of the Himalayas receive more rainfall *e.g.* Simla 1520mm, Nainital 2000mm and Drazeeling 3150 mm whereas the places away from the Himalayan foothills receive relatively low rainfall *e.g.* Patna 1000 mm, Allahabad 1050 mm and Delhi 650 mm.

(c) If the mountains are of moderate height, the maximum rainfall does not occur at their tops rather it occurs on the other side.

(d) The windward slopes of the mountains at the time of rainfall are characterized by cumulus clouds while leeward slopes have stratus clouds.

(e) The amount of rainfall increases with increasing height along the windward slopes of the mountains upto a certain height beyond which the amount of rainfall decreases with increasing height because of marked decrease in the moisture content of the air. This situation is called inversion of rainfall. The height of the mountains beyond which the amount of rainfall decreases upward is called maximum rainfall line which varies spatially depending on the location of mountains, their distance from the sea, moisture content in the air, mountain slope, season etc. Maximum rainfall line is at 24,000 feet (7,000 m) at the equator, at 12,000 feet (3,600 m) in the Himalayas, at 21,000 feet (6,300m) in the Alps, at 18,000 feet (5,400 m) during summer and at 12,000 feet (3,600m) during winter in the Pyrenees mountains etc.

(f) Orographic rainfall may occur in any season. Unlike other types of rainfall it is more widespread and of long duration.

(g) It may be pointed out that orographic rainfall is induced not only because of lifting of moist air due to mountain barrier but convective and cyclonic mechanisms also help in the process of orographic rainfall. For example, in warm regions valleys are heated during daytime and hence winds are also heated and ascend along the hillslopes in the form of convection currents and yield rainfall after being saturated. Some times, forward moving cyclones are also forced to ascend along the hillslope due to obstructions offered by mountain barriers.

(3) Cyclonic or Frontal Rainfall

Cyclonic or frontal rainfall occurs due to ascending of moist air and adiabatic cooling caused by convergence of two extensive air masses. The mechanism of cyclonic precipitation is of two types on the basis of two types of cyclones viz. temperate cyclones and tropical cyclones. Rainfall associated with temperate cyclones occurs when two extensive air masses of entirely different physical properties (warm and cold air masses) converge. When two contrasting air masses (cold polar air mass and warm westerly air mass) coming from opposite directions converge along a line, a front is formed. The warm wind is lifted upward along this front where as cold air being heavier settles downward. Such cyclonic fronts are created in temperate regions where cold polar winds and warm westerlies converge. The warm air lying over cold air is cooled and gets saturated and condensation begins around hygroscopic nuclei. It may be pointed out that lifting of warm air along cyclonic front is not vertical like convective currents rather it is oblique. Since the lifting of warm air along the warm front of temperate cyclone is slow and gradual and hence the process of condensation is also slow and gradual, with the result precipitation occurs in the form of drizzles but continues for longer duration. Thus, the precipitation associated with warm front is widespread and of long duration. On the other hand, the precipitation associated with cold fronts is always in the form of thunder showers but is of very short duration. Sometimes, the precipitation occurs in the form of snowfall and hailstorms. This is because of the fact that lifting of warm air along cold front occurs quickly as cold air pushes warm air upward with great force. Most of the rains of temperate regions are received through cyclones.

In tropical regions two extensive air masses of similar physical properties converge to form tropical cyclones wherein lifting of air is almost vertical and is very often associated with convection. It may be pointed out that convergence mechanism provides initial trigger effect to the upward movement of convectively unstable air which if full of moisture becomes saturated and yields heavy showers characterized by lightning and thunder. Tropical cyclones, regionally called as typhoons, hurricanes, tornadoes etc. yield heavy downpour in China, Japan, South-East Asia, Bangladesh, India, USA etc.

Global Distribution of Rainfall

Rainfall is highly correlated with air temperature and atmospheric humidity while humidity is closely related with temperature through the process of evaporation. The regions having high temperature and abundance of surface water for evaporation receive higher amount of annual rainfall. Equatorial regions are typical example of such situation. Subtropical regions are also characterized by above conditions but the western parts of the continents receive least rainfall because there are anticyclonic conditions due to descent of air. Middle latitudes also have favourable conditions for sufficient rain- fall but polar areas receive their precipitation in the form of snowfall instead of rainfall. Before attempting to describe world distribution of rainfall it is necessary to discuss certain facts related to rainfall distribution *e.g.* total amount of annual rainfall, seasonal distribution, and variability of rainfall.

Mean annual rainfall for the whole globe is 970 mm but this mean annual amount is unevenly distributed on the earth's surface. Some places receive less than 100 mm of mean annual rainfall (for example, tropical hot deserts like Sahara, Thar, Acatama, Kalahari etc.) while some places receive more than 12000 mm of annual rainfall (*e.g.* Cherrapunji of India). Not only this, there is much temporal variation of annual rainfall in a particular area. Most of the annual amount of rainfall is received during a few months of the year while most of the months either remain dry or receive little rainfall. For example, 12000 mm of rainfall at Cherrapunji is received only in 159 days. The equatorial regions receive rainfall throughout the year but other areas are characterized by seasonal rainfall. For example, more than 80 per cent of annual rainfall, in India is received during 3 wet summer monsoon months (July, August and September). On the other hand, the Mediterranean regions receive most of their annual rainfall during winter months while summer season remains dry.

Zonal Distribution of Rainfall

It is true that the cooling of ascending air is a prerequisite condition for the occurrence of rainfall. The air is lifted through thermal convective mechanism, convergence of two extensive air masses and obstruction of mountain barriers. If the relative importance of these three factors in different areas of the world is taken into account, it appears that air is generally lifted due to convergence of two extensive air masses in most parts of the world. Convergence of air masses is directly related with air temperature and air pressure. There are two major convergence zones of air masses *i.e.*, trade winds converge along the equatorial low pressure belt and westerlies and polar winds converge along high latitude low pressure (600-650 latitudes in both the hemispheres). On the other hand, winds descend near subtropical high pressure belt (30°-350 latitudes in both the hemispheres) and diverge in opposite directions and form anticyclones which introduce dry weather. Since the convergence of air masses is in zonal form and hence rainfall distribution is also found in zonal pattern. Besides, mountain barriers and land and water (continents and oceans) also influence world distribution of rainfall. Since air moisture depends upon temperature and horizontal distribution of temperature is found in zonal patterns and hence rainfall distribution is also characterized by zonal pattern. Based on above considerations, 6 major zones of rainfall distribution are identified on the earth's surface.

(1) *Equatorial zone of maximum rainfall*—This zone extends upto 10° latitudes on either side of the equator and falls within intertropical convergence characterized by warm and moist air masses. The mean annual rainfall ranges between 1750 mm and 2000 mm. Most of the rains are received through convectional rainfall accompanied by lightning and cloud thunder. There is daily rainfall in the afternoon. The rainfall intensity is very high as it occurs in the form of heavy showers. The clouds are cleared within short period and sky becomes cloudless in the late evening.

(2) *Trade wind rainfall* zone extends between 10°-20° latitudes in both the hemispheres and is characterized by north-east and south-east trade winds. These winds yield rainfall in the eastern parts of the continents because they come from over the oceans and hence pick up sufficient moisture but as they move westward in the continents they become dry and thus the western parts of the continents become extremely dry and deserts. The monsoon regions located in this zone receive much rainfall. Summers receive most of the mean annual rainfall.

(3) *Subtropical zone of minimum rainfall* extends between 20° and 30° latitudes in both the hemispheres, where descending air from above induces high pressure and winds diverge in opposite directions at the ground surface, with the result anti-cyclones are formed. This condition is not conducive for rainfall and hence dry conditions prevail over large areas. Mean annual rainfall is 900mm. It may be pointed out that all the tropical

hot deserts are located in this zone where mean annual rainfall is below 250mm. The average annual rainfall becomes, for the whole zone, higher (900mm) than the average value for the deserts because the eastern parts of the continents receive more rainfall from relatively moist trade winds which come from over the oceans. Most of annual rainfall occurs during summer months while winter season is dry.

(4) *Mediterranean rainfall* zone extends between 30°-40° latitudes in both the hemispheres where rainfall occurs through westerlies and cyclones during winter season while summers remain dry because this zone comes under the influence of trade winds due to northward shifting of wind and pressure belt during northern summer (summer solstice). Mean annual rainfall is 1000mm.

(5) *Mid-latitudinal zone of high rainfall* extends between 40°-50° latitudes in both the hemispheres where rainfall occurs through westerlies and temperate cyclones. Mean annual rainfall ranges between 1000m and 1250 mm. The western pans of the continents receive more rainfall. It decreases from the western coastal areas inland. Southern hemisphere records more rainfall than northern hemisphere because of dominance of oceans in the former. Winter season receives maximum precipitation through temperate cyclones. The precipitation is of long duration but occurs in the form of light showers.

(6) *Polar zone of low precipitation*—Precipitation decreases from 60° latitude poleward in both the hemispheres. Mean annual precipitation becomes only 250mm beyond 75° latitude. Most of the precipitation occurs in the form of snowfall.

Patterson has divided the globe into 15 rain-fall zones wherein the northern and southern hemispheres account for 7 zones each and the remaining zone is on either side of the equator. These rainfall zones have been identified main^ on the basis of seasonal behaviour of rainfall.

	latitudes
(i) Rains throughout the year	7°N to 7°S
(ii) Summer rains, winter dry	7°-16°
(iii) Light summer rains	16°-200
(iv) All seasons dry, minimum rains	20°-300
(v) Light winter rains	30°-35°
(vi) Summers dry, winter rains	35°-45°
(vii) All seasons rains, maximum in summers	40°-70°
(viii) All seasons scanty precipitation mostly snowfall	70°-900

Rainfall Regime

Precipitation or rainfall regime refers to seasonal behaviour and variation of rainfall. Haurwitz and Austin have identified 6 rainfall regimes.

(1) *Equatorial rainfall regime* is characterized by rainfall in all seasons but there are two maxima in March (vernal equinox) and September (autumnal equinox). Thermal convection air currents generated by intense insolational heating of the ground surface account for most of the rains. Besides, convergence of trade winds also causes cyclonic rains. This zone extends between 10°N and 10°S latitudes. At the outer limit of this zone there is only one rainfall maximum. The rainfall is accompanied by lightning and thunder and occurs in the form of heavy showers but is of short duration.

(2) *Tropical rainfall regime* has one rainfall maximum and one minimum in a year. In the northern hemisphere the eastern parts of the continents receive maximum and minimum rainfall in the months of July and December respectively whereas the western pans of the continents get maximum rainfall in December and minimum in July.

(3) *Monsoon rainfall regime* is characterized by maximum rainfall in July and August (northern hemisphere). Thus, there is summer maximum and winter minimum. Summer-rainfall is received through south-west arid south wet monsoon winds associated with tropical atmospheric disturbance (cyclones). Most of the rains are orographic and, cyclonic in origin.

(4) *Mediterranean rainfall regime* receives maximum rainfall during winter season because the zone of this regime comes under the domain of prevailing westerlies during winters due to southward shifting of pressure and wind belts. Summer is a dry season.

(5) *Continental rainfall regime* is characterized by maximum precipitation in summers when convective mechanism due to insolational heating of the ground surface is maximum. Winters are dry because of the prevalence of anticyclonic conditions. This regime is found in the interior of the continents.

(6) *Maritime rainfall regime*—Temperate areas record maximum precipitation in winter over the oceans and adjoining coastal areas due to maximum cyclonic activity. This regime is found along the western margins of the continents in middle latitudes.

COOLING OF AIR AND ADIABATIC CHANGE OF TEMPERATURE

Temperature decreases with increasing height at the rate of 6.5°C per 1000m or 3.6°F per 1000 feet. This rate of decrease of temperature with increasing height is called normal lapse rate. A definite ascending air with given volume and temperature expands due to decrease in pressure and thus cools. For example, an air with the volume of one cubic foot and air pressure of 1016 mb at sea level if rises to the height of 17,500 feet, its volume is doubled because of expansion. On the other hand, a descending air contracts and thus its volume decreases but its temperature increases. It is apparent that there is change in temperature of air due to ascent or descent but without addition or substraction of heat. Such type of change of temperature of air due to contraction or expansion of air is called *adiabatic change of temperature*. Fig. 2.5 depicts the mechanism of cooling of ascending air due to expansion of volume (adiabatic change of temperature).

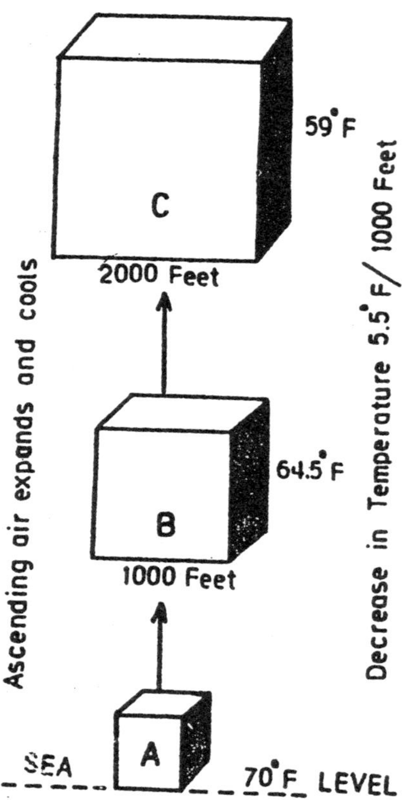

Fig. 2.5 : Expansion and cooling of air due to lifting and adiabatic change of temperature.

Adiabatic change of temperature is of two types viz. (i) *dry adiabatic change* and (ii) *moist adiabatic change*. The temperature of unsaturated ascending air decreases who increasing height at the rate of 5.5°F per 1000 feet or 10°C per 1000m. This type of change of temperature of unsaturated ascending or descending air is called dry adiabatic rate. It may be pointed out that if an air descends its temperature increases at the above mentioned rate. The rate of decrease of temperature of an ascending air beyond condensation level is lowered due to addition of latent heat of condensation to the air.

This is called moist adiabatic rate wherein temperature of an ascending air beyond condensation level decreases (and hence the air cools) at the rate of 3°F per 1000 feet or 6°C per 1000m. This is also called as retarded adiabatic rate.

STABILITY AND INSTABILITY OF THE ATMOSPHERE

Different forms of precipitation (dew, fog, rainfall, frost, snowfall, hailstorm etc.) depend on stability and instability of the atmosphere. The air without vertical movement is called stable air while unstable air undergoes vertical movement (both upward and downward). An airmass ascends and becomes unstable when it becomes warmer than the surrounding airmass while descending airmass becomes stable. The stability and instability depend on the relationships between 'normal lapse rate' and 'adiabatic change of temperature'. Adiabatic rate is always constant whereas normal lapse rate of air temperature changes. When the normal lapse rate is higher than dry adiabatic rate, the air being wanner rises and becomes unstable. On the other hand, when the normal lapse rate of temperature is lower than dry adiabatic rate, the air being cold descends and becomes stable.

Stability

When dry adiabatic lapse rate of an ascending dry air is higher than the normal lapse rate and if it is not saturated and does not attain dew point it becomes colder than surrounding air at certain height with the result it becomes heavier and descends. This process causes stability of atmospheric circulation due to which vertical circulation of air is resisted. For example, at ground surface if the temperature of a parcel of air is 40°C, the dry adiabatic lapse rate and normal (environmental) lapse rate are 10°C per 1000m and 6.5°C per 1000 m respectively, then at the height of one kilometre (or 1000 m) from the ground surface the temperature of the ascending air would be 30°C (40°–10° = 30°C) while the temperature of surrounding air at that height would be 33.5°C (40°-6.5° = 33.5°C). Thus, the ascending air being colder than surrounding air would descend and atmospheric stability is caused. Such air (descending) is called to be in *state equilibrium*. Some times, the normal lapse rate in a certain layer of the atmosphere is about 4.6°C per 1000 metres. In such conditions if the normal lapse rate is less than wet adiabatic lapse rate even at condensation point, further vertical motion of air is stopped and thus such air is said to be absolutely stable and such atmospheric condition is called absolute stability.

Instability

When normal lapse rate is greater than dry adiabatic lapse rate of ascending parcel of air the rising air continues to rise upward and expand and thus becomes unstable and is in unstable equilibrium. In other words, atmospheric instability is caused when the rate of cooling of rising air (dry adiabatic lapse rate) is lower than the normal lapse rate. For example, if the temperature of a certain parcel of air at ground surface is 40°C, the dry adiabatic and normal lapse rates are 10°C and 11°C per 1000m respectively, then the temperature of ascending air at the height of 1000m (one kilometre) would be 30°C (40°-10° = 30°C) while the temperature of the atmosphere at that height would be 29°C (400-110C = 29°C). Thus, the rising air being warmer (30°C) than the surrounding air (29°C) continues to rise and expand to cause atmospheric instability. If the wet adiabatic lapse rate is also less than normal lapse rate, the rising air further continues to rise upward. Such state of continued upward movement of air is called absolute instability. When the ascending parcel of air reaches such height that its temperature equals the temperature of surrounding air, its further upward movement is stopped. Such air is said to be in the state of neutral equilibrium.

Mechanical Instability

Some times, there are abnormal conditions when normal lapse rate is exceptionally very high (15° to 35°C per 1000 metres). In such condition the upper layers of the atmosphere become exceptionally cold and denser than the underlying layers, with the result cold and denser upper layers automatically descend. Such situation is called mechanical instability which helps in the formation of tornadoes.

Conditional Instability

When a parcel of air is forced to move upward, it cools at dry adiabatic lapse rate (10°C per 1000m, or 5.5°F per 1000 feet) whereas normal lapse rate is 6.5°C per 1000m. After rising to certain height the air becomes saturated and latent heat of condensation is added to the rising air so the rising air cools at wet adiabatic lapse rate (5°C per 1000m) whereas the normal lapse rate (6.5°C per 1000m) is greater than it. Consequently, the air becomes wanner than the surrounding air and hence rises upward automatically. This is called conditional instability because the air is initially forced to move upward but rises automatically due to its own properties after condensation point is reached. For example, if a parcel of air with 35°C temperature is initially forced to rise upto the height of 1000m, its temperature decreases to 25°C (35°C-10°C, dry adiabatic rate = 25°C) whereas the temperature of surrounding layers of air at the height of 1000m would be 28.5°C (350C-6.50C, normal lapse rate) and thus the rising air becomes colder by 3.5°C than the surrounding air. If the rising air gets saturated at this temperature (25°C), the latent heat of condensation returns back to the rising air and hence it cools at the wet adiabatic lapse rate (5°C per 1000m).

Thus, the rising air becomes warmer and unstable, Conditional instability may occur only when the normal lapse rate ranges between dry adiabatic and wet adiabatic lapse rates. In other words, conditional instability occurs when normal lapse rate is greater than dry adiabatic lapse rate but less than wet adiabatic lapse rate.

EVAPORATION

Evaporation is the process by which water is changed from the liquid to the gaseous form, a process of molecular transfer, and so incorporated into the atmosphere. The rate and amount of evaporation in many ways is as important to the climatologist as is the amount of rainfall. A high evaporation rate reduces the effectiveness of the rainfall; in northern Sri Lanka, for example, the 'Dry Zone' receives about 130 cm (50 in) of rain per annum, an amount which would be regarded as distinctly heavy in middle latitudes. The *effectiveness of precipitation* is taken to be the total rainfall minus the total possible evaporation.

Evaporation is measured in a variety of ways, though none of them is entirely satisfactory. The Piche evaporimeter is commonly used, in which water in a tube is allowed to evaporate from a piece of porous paper, and the loss in a certain time is measured on a graduated scale along the tube. Some evaporation statistics are calculated from measurements taken of the water-level in large open tanks. This is more correctly known as *potential evaporation*, since it depends on the availability of a constant supply of water.

The *rate of evaporation* is mainly a function of relative humidity, of absorbed radiation and of air movement, but it depends, too, on the nature of the surface. The loss from bare soil is very rapid, but where the surface is covered with a loose tilth (as in dry-farming practice) it is much reduced. A plant cover may protect the ground itself against direct evaporation by the shade it provides, but the loss by transpiration, plus direct evaporation (a joint effect sometimes called *evapotranspiration*), may be very great. Experiments were carried out by the Fylde Water Board at the Stocks Reservoir, Slaidburn, Yorkshire, where the close planting of spruce has taken place. A total of 98-4 cm (38-75 in) of rain fell there in one year, but as a result of the interception of rainfall by the trees and of evaporation from the foliage, only 61 cm (24 in) reached the ground, and of that only 27-3 cm (10-75 in) became available for water supply. In other terms, on 600 hectares (1500 acres) of catchment the planting of spruce reduced the available water supply by 4-5 million litres (i million gallons) a day. It must be admitted, however, that not all authorities accept the validity of this experiment and the deductions from it.

The highest potential evaporation records come from the Tradewind deserts, where the effects of high temperatures, strong fresh winds and a bare sandy or rocky surface are combined. Atbara and Khartoum in the Sudan have mean annual potential evaporation figures of 625 cm (246 in) and 541 cm (213 in) respectively, Helwan in Egypt of 239 cm (94 in). There are considerable seasonal differences; the mean evaporation at Helwan in June is 33 cm (13 in), but in January and December it is only 8-9 cm (3-5 in). Vast quantities of water are evaporated from the oceans in subtropical latitudes, are injected into the atmosphere, condense as clouds, and move Equatorwards.

Evaporation rates are low in the equatorial belt, where they only amount to 5-8 cm (2-3 in) a month. They are low, too, in cool middle latitudes; the mean annual rate of evaporation for London is only 46 cm (18 in) (Fig. 2.6).

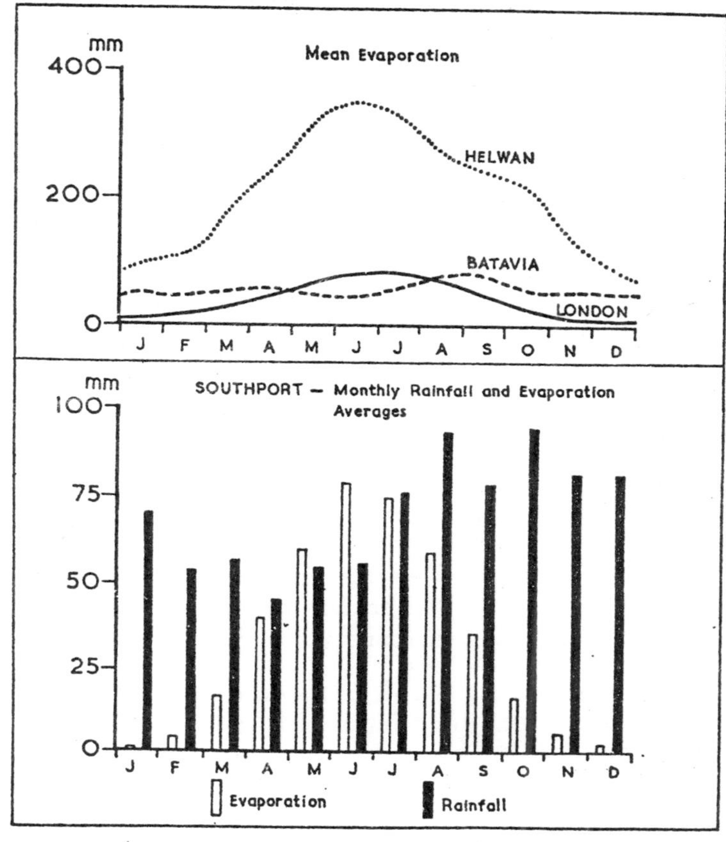

Fig. 2.6 : Evaporation.

Causes of Condensation

Condensation has been defined as the formation of water-droplets when air has been cooled to and beyond its dew-point. There are several ways in which air may be so cooled, including direct radiation from the surface of the earth during a clear night the horizontal movement of warm air over a cold surface, the mixing along the margins of two air currents of markedly different temperature, the movement of air from warmer to cooler latitudes, and, by far the most important, by ascent. Each form of cooling may produce condensation of different degrees and with different results.

It has been shown by experiment that completely pure air can be cooled, under laboratory conditions, to temperatures well below the dew-point without condensation taking place. It is necessary for some kind of nuclei to be present on which the droplets can form. These nuclei include particles of dust and smoke, salt from the ocean, pollen, and even negative ions (atoms carrying a negative electrical charge produced by the passage of radiation through the atmosphere).

The condensed droplets are only about 0.05 mm in diameter when they form, and are so minute that they float in the air as fog or clouds. Larger drops form on leaves and grass as dew, or if the temperature is below freezing-point as hoar-frost. When droplets coalesce in the air to a certain critical size, they may fall to earth as one or other of the forms of precipitation (rain, snow, hail or sleet). The formation of raindrops is, however, much more complex than this simple account would imply. Indeed, it is not known how droplets actually coalesce, though various theories propose electrical attraction, supercooling followed by freezing into ice particles (which form nuclei for condensation), and turbulence causing coalescence by collision.

FOG

General Characteristics

Fog is special type of thin cloud consisting of microscopically small water droplets which are kept in suspension in the air near the ground surface and reduces horizontal visibility. According to Byers fog is defined 'as almost microscopically small water drops suspended in the atmosphere and reducing the horizontal visibility to less than one kilometre'. It may be pointed out that clouds are formed due to ascent, expansion and cooling of air while fogs are formed due to radiation, conduction and mixing of warm and cold air masses near the earth's surface.

Fog is formed when the moist air (with relative humidity above 97 per cent) becomes saturated, reaches its dew point and further cools so that water vapour is condensed around dust particles, smokes etc. to form tiny water droplets which being very light are suspended in the air. These microscopically tiny water droplets form smoky clouds which fly with the winds. Invisibility increases as the fog becomes dense and there is darkness when the fog becomes extremely dense. A light fog, called as mist, is that when visibility is restricted to 2 kilometres. Fog is generally associated with inversion of temperature and occurs in the morning hours but some times also continues till noon. Fog occurs during winters in subtropical regions but it occurs in all seasons in the regions beyond 35° latitudes. Generally, fog looks whitish in colour but over large cities and industrial areas it looks dirty yellow or gray because of mixing of smoke, dust and fly ash.

Classification of Fogs

Fogs have been variously classified by different scientists on various bases. H.C. Willett (1928) classified fogs in two broad categories on the basis of physical processes of their formation *e.g.* (1) air mass fog and (2) frontal fog.

H.R. Byers (1944) modified the classification of fogs as suggested by Willett and presented the modified form as follows :

(A) Air Mass Fogs

(1) Advective types :

(a) Fogs due to the transport of warm air over a cold surface

(i) land and sea breeze fog

(ii) sea fog

(iii) tropical fog

(b) Fogs due to transport of cold air over a warm surface

(i) steam fog (arctic sea smoke)

(2) Radiation types :

(a) ground fog

(b) high inversion fog

(3) Advection-radiation fog

(4) Upslope fog

(B) Frontal Fogs

(1) Pre-frontal (warm front) fog

(2) Post-frontal (coid front) fog

(3) Front passage fog

Critchfield has classified fogs in two broad categories on the basis of evaporation and cooling processes which are responsible for the origin of fogs—

(1) Fogs resulting from evaporation

(i) steam fog

(ii) frontal fog

(2) Fogs resulting from cooling

(i) radiation fog

(ii) advection fog

(iii) upslope fog

(iv) mixing fog

(v) barometric fog

Fogs are classified in 4 types on the basis of visibility :

(i) light fog (visibility upto 1100 metres)

(ii) moderate fog (visibility 1100 m-550 m)

(iii) dense fog (550 m-300 m)

(iv) indense dense fog (less than 300m).

Radiation fog is formed when warm and moist air lies over cold ground surface. Due to this situation overlying warm and moist air cools and thus dew point is reached, with the result condensation of water vapour around hygroscopic nuclei (dust particles and smokes) forms numerous tiny water droplets and thus fog is originated. The occurrence of radiation fog requires certain conditions *e.g.* long and cool winter nights, cloudless sky, sufficient amount of moisture in the air, very weak air motion (light wind having speed of 3 to 5 km per hour) and ground inversion of temperature. Fog is formed at the ground surface during nights and thickness upward. It disappears in the morning with the sunrise because the water droplets are evaporated due to rise in air temperature. Radiation inversion occurs only on land surface. Radiation fog is more common over large cities and surrounding areas because of abundance of hygroscopic nuclei. When fog is combined with sulphur dioxide it becomes poisonous and causes human deaths. Such fog is called *urban smog*. Sometimes, fogs also occur at higher elevation due to upper air inversion of temperature. Such fog is called *high inversion fog*.

Advectional Radiation Fog

The fog formed due to mixing of warm moist air and cold air due to arrival of warm and moist air over cold ground surface is called advectional radiation fog. There are certain necessary conditions for the origin of advection radiation fog. *e.g.* (i) horizontal movement of air, (ii) greater contrast between air temperature and the temperature of ground surface, (iii) moist air or say high relative humidity in the air, (iv) stable stratification in the atmosphere etc. The warm and moist air when blows over cold surface (either land surface or sea surface) is cooled from below so that dew point is reached and condensation of water vapour occurs around hygroscopic nuclei and thus microscopic water droplets are formed to create fogs. Advection fogs are generally originated during winters on land surfaces and during summers on sea surfaces because lands are relatively colder than seas and oceans during winters. Dense advectional fogs are also formed where cold and warm ocean currents coverage. For example, dense fogs are formed near New-foundland due to convergence of cool Labrador and warm Gulf Stream ocean currents. Similarly, dense fogs are formed near Japanese coast due to convergence of cool Kurile and warm Kuroshio currents.

The fogs occurring over sea surfaces are called sea fogs which are generally formed near the coastal areas frequented by cold ocean currents. For example, cool California current (along the Californian coast), cool Peru current (along the Peruvian coast), cool Benguela current (along the western coast of South Africa) and cool Western Australian current cause sea fogs. The sea breezes transport sea fogs inland upto the distance of 400 km.

Steam fogs are in fact advectional fogs which are formed when cold air moves from land" over oceanic surface and there is evaporation of large quantity of moisture from water surface to saturate the overlying cold air. It is interesting to note that in such situation overlying cold air is warmed from below and thus water vapour due to evaporation of warm water surface rises upward and condenses after meeting cold air above to form fog. The vapour panicles above the water surface look as if steam is coming up from the water. This is why such fogs are called steam fogs. They are also called evaporation fogs.

Upslope or hill fogs originate when continental warm and moist air rises upslope along the hillslopes because the rising air is saturated due to cooling and condensation of moisture around hygroscopic nuclei and forms fogs which cover the lower segments of hillslopes. It may be mentioned that such fogs are formed due to adiabatic expansion and cooling of moist air. Such fogs are very common on the hillslopes in temperate regions. They may occur in any season.

Frontal Fogs

Fronts are formed when two contrasting airmasses (warm and cold airmasses) converge along a line. Warm air is pushed upward by cold air and hence overlying warm air is cooled from below due to underlying cold air and fogs originate after condensation. Such fogs are formed in temperate regions. In fact, the saturation and condensation of overlying warm and moist air causes rainfall which saturates the underlying cold surface air and thus fog is produced.

World Distribution of Fogs

If the world distribution of fogs is considered it becomes apparent that they are more widespread over oceans. The oceanic fogs are generally advectional in origin because relatively warm air coming from over the land surfaces produces fog over relatively colder oceanic surfaces. Fog areas are positively related to cool ocean currents. The eastern portions of the oceans and western continental margins are characterized by most fogs because these areas are frequented by cool ocean currents (*e.g.* California current, Peru current, Benguela current, Canary current, Western Australia current etc.). Frontal fogs are produced in the high latitudes due to convergence of cold polar air mass and warm westerlies mainly in the eastern north Atlantic Ocean and western coasts of north-west Europe and in the eastern north Pacific Ocean near Aleutian Islands and Alaska coast. Radiation fogs are produced in low latitudes on continental areas during winter season. Dense fogs are formed at the places where cold and warm ocean currents converge *e.g.*, near New-foundland and Grand Bmks (due to convergence of cold Labrador current and warm Gulf Stream) and near Japanese coasts (due to convergence of cold Kurile current and warm Kuroshio current).

Important foggy areas of the USA are California coast. New England outer coast. Northern Pacific coast line, Applachian valleys. Pacific coast valleys, Middle Atlantic coast, Great Lakes, Southern Atlantic and Gulf coastal waters, Ohio, Missouri, Great Plains and upper Mississippi valley.

Effects of Fogs

Fogs effectively hinder sea navigation, land and air transport systems. Dense fogs cause severe accidents on high-speed highways involving collision of trucks, buses, cars etc. Landing and take-off of aircrafts are delayed by dense fogs causing economic loss to airlines and inconvenience to stranded passangers.

Dense fogs are navigational hazards in the seas as ships and huge supertankers carrying oil some times collide resulting into spilling of huge quantity of oil causing enormous oil slicks on sea water which cause ecological disaster. Vehicular traffic comes to grinding halt in the event of dense fogs. Some times, fogs are so dense and visibility is so reduced that walking by human beings becomes impossible.

When fogs are polluted through sulphur coming out of the chimneys of the mills, they become poisonous and health hazards. For example, killer poisonous smog was formed due to mixing of dense fog with smokes and sulfide fumes coming out from the chimneys in the Donora Valley of Pennsylvania (USA) on 26 October, 1948. This toxic fog (smog) caused 20 human deaths and 43 per cent inhabitants fell ill due to cough and respiratory problem.

Similar poisonous fog was produced due to mixing of sulphur dioxide coming out of the factories of zinc smelters and sulphuric acid in the Meuse Valley of Belgium in the month of December, 1930. This killer fog claimed 63 human lives due to obstruction in respiratory system. The poisonous fog of December, 1952 claimed 4000 human lives in London.

Sometimes, fogs are also economically beneficial to human society. For example, they protect tea and coffee plants from scorching sunlight on the hillslopes. Mocha coffee on the Yemen hills (Arabia) is highly benefitted from fogs.

It is sometimes possible for drops to form and then continue to exist in the liquid state, even when the temperature is well below freezing, as long as the air is undisturbed. This phenomenon is known as *supercooling*, and has a number of important meteorological results. One practical effect of great importance is the accretion of ice on aircraft. If an aircraft passes into a cloud consisting of large drops of supercooled water, with the air temperature at or below freezing-point, a considerable thickness of clear ice may form as each drop freezes on impact with the leading edges of the wings. A similar phenomenon is known as *glazed frost* or as 'silver thaw' in America. The supercooled water freezes on branches, telegraph wires (which may be brought down by the weight), and road surfaces (which become treacherous). Rime can also form on aircraft. Incidentally, another cause of icing on an aircraft is when it is flying in a layer of air of temperature below freezing-point, but not in a cloud; if heavy rain falls from a cloud above, such as might occur in a warm front over a cold air-mass, it will immediately, freeze into clear ice on the aircraft. The provision of icing warnings has become an important function of forecasting at the meteorological stations attached to airports and Service airfields.

Dew

Dew occurs when condensation takes place on the surface of bodies which have cooled by nocturnal radiation to a temperature below the dew-point of the layer of air resting on the earth's surface. This is particularly likely to occur after a warm day, during which there has been much evaporation to increase the vapour concentration of the air. The evening must be calm, since wind mixes the layers of air and does not allow one layer to remain sufficiently long in contact with the earth for cooling to take place, and the sky must be clear since absence of cloud allows radiation of heat to take place readily.

The dews of spring and early summer are probably mainly derived from water-vapour in the atmosphere. However, experiments have shown that, particularly in autumn when the earth is warm, the moisture deposited as dew can come from the ground itself, both directly and transpired by plants (*guttation dew*). It is condensed on grass and other bodies which have cooled by radiation—undersides of objects left overnight on a lawn are often wet with dew in the morning.

A practical application of dew formation is shown by the *dewmounds* which are a common feature in the drier parts of Israel. A mound of earth is covered with flat stones on which copious dew condenses, trickling between the stones into the earth, which is kept moist. Citrus fruit trees grow out of the mounds, and are thus supplied with water.

Mention may be made of a *dew-pond*, although this seems to be something of a misnomer. Hollows were dug in the South Downs and lined with puddled clay and straw, in order to provide sheep and cattle with water. Most of the water is derived from rainfall, either directly on to the surface of the pond or by trickling down the sides of the hollows; some comes from the condensation of sea-mists from the English Channel, but only a small amount (about 25 mm annually) derives actually from dew.

Hoar-frost forms when the dew-point is below freezing-point, so that the vapour is deposited directly in the form of tiny ice spicules, without passing first through the liquid state.

Mist and Fog

Fog, another result of condensation, is notable mainly for the resulting reduction in visibility, and in this respect it is an important meteorological element which forms a major handicap to transport by land, sea and air. The term *fog* is used (under the International Meteorological Organization Code) when the visibility is less than 1 km, *mist* when the visibility extends from 1 to 2 km. In Great Britain, however, 'thick fog' implies a visibility of less than 200 m. The term *haze* is reserved for impaired visibility of 1 to 2 km as a result of dust or smoke particles. The categories of 'poor', 'moderate', 'good' and 'very good' visibility are defined on an international scale in terms of the range of vision from an observing station.

The main types of fog, according to the causes responsible for them, are:

(i) Radiation fog,

(ii) Advection fog,

(iii) Frontal fog,

(iv) Steam fog and

(v) Hill fog.

(i) *Radiation fog* : This type owes its formation to an amplification of the conditions which produce dew. During a period of calm, clear weather in Britain, particularly during spring and autumn, the surface of the earth is rapidly cooled by radiation, and so the layer of air resting upon it is also cooled. In hilly country the cooled dense air flows by gravity into hollows, and as it comes into contact with the cold earth, condensation produces a shallow horizontal layer of white radiation fog. Often the layer of fog survives well after sunrise;

in the English Lake District, particularly in early summer, one can stand on a hillside in the morning sunlight while the valley below is still mist-filled, but as the sun rises in the sky the fog is dissipated.

Under cold anticyclonic conditions in late autumn and winter, the radiation fog may be thicker and more persistent. The sun's rays are then less powerful, and the fog may linger until the calm weather of the anticyclone is interrupted. It tends to form over valley floors, where the air is moist (particularly when there is a river), and where the chimneys of a large town pour into the atmosphere quantities of soot which act as nuclei. Such are the 'pea-soup fogs' or 'smogs' formerly common in London, Merseyside and most big urban areas. Sulphur dioxide poured into the atmosphere from burning coal combines with the moisture present to form sulphuric acid, thus giving these urban fogs their acrid flavour.

One of the worst fogs ever experienced was in London in December 1952. An anticyclone over the Thames valley formed a deep pool of cold stagnant air, with a much warmer layer above, creating an impenetrable inversion and resulting in an enormous accumulation of soot. Between 6 and 8 December visibility was down to a few metres. The smog resulted in 4000 deaths from bronchitis and pneumonia, and a vast bill for cleaning curtains and clothing. This state of affairs has led to campaigns for smoke abatement, the creation of smokeless zones, and the use of smokeless fuels. The National Smoke Abatement Society publishes a journal, *Smokeless Air*. Improvements can be effected, as shown by Pittsburgh, one of the dirtiest towns in the U.S.A. for many years, which achieved a reduction of atmospheric pollution of over 90 per cent in six years.

(ii) *Advection fog :* When a warm, moist air current is cooled as it moves horizontally over a cold land or sea surface, an advection fog may be formed. The term advection is used of a horizontal transfer of air, as compared with convection when the change is vertical.

Where the hot deserts reach the west coasts of continents occur what are called 'cold-water coasts'. These are due to the presence near the coast of cool currents flowing equatorward, accentuated by the upwelling of cold deep water. Onshore fogs, formed by condensation over the cool water, are blown for a few km inland by the sea breezes, but they gradually dissipate over the much warmer land. The Golden Gate at the entrance to San Francisco harbour experiences dense fogs of this nature on an average of forty days in the year.

A striking example of an advection fog forming over the sea is in the neighbourhood of the Grand Banks of Newfoundland. When air from over the Gulf Stream, moving in a northerly direction, crosses the line of the 'cold wall', it passes over the waters of the Labrador Current, which are some 8° to 11°C cooler, since the Labrador is bringing melt-water from disintegrating pack-ice, while the Gulf Stream 'transports' high temperatures from much farther south. A thin layer of dense fog persists for many days; in the Strait of Belle Isle it occurs on an average on more than 100 days in the year, and off Newfoundland on more than 70 days. These fogs are common along the coast of north-eastern America and cause much delay to shipping. In July 1959, for example, the Queen Elizabeth was in collision with a freighter in thick fog soon after leaving New York.

(iii) *Frontal fog* : Short-lived fog, really a very thick, fine drizzle, is sometimes associated with the passage of the warm front of a depression. The warm rain falls into the cold underlying air near the ground, and if there is a marked temperature difference gloomy foggy conditions may prevail for a time.

(iv) *'Steam fog'* : This type is comparatively rare, but it is included for its meteorological interest. It is formed when cold air passes over a much warmer water surface, so that the water appears to 'steam'. In high latitudes it is known as *ice-fog*, where the moisture in the air is converted into ice crystals, but it rarely lasts for any length of time. A distinction is sometimes made between steam fog, which develops over fresh water, and 'Arctic Smoke' over salt water.

(v) *Hill fog* : This is simply low sheet-cloud, which may envelop, for example, the hills of western Britain when a moist air-stream is moving inland. Mountainous districts in middle latitudes are liable to spells of hill fog at any time of year during unsettled weather.

Rime

When a fog composed of supercooled droplets is driven by a slight wind against prominent objects such as telegraph poles, wires and trees the drops freeze on to them as rime. The temperature both of the droplets and of the objects is below freezingpoint. The ice particles imprison some air, so that the rime has a white, opaque appearance. This is a common feature on the mountains of Britain, and the climber in winter may find his progress impeded by the covering of 'frost-feathers' on the rocks.

CLOUDS

Clouds are defined as aggregates of innumerable tiny water droplets, ice particles or mixture of both in the air generally above the ground surface. Clouds are formed due to condensation of water vapour around hygroscopic nuclei caused by cooling due to lifting of air generally known as adiabatic cooling.

Meteorologically clouds are very significant because all forms of precipitation occur from them. It may be mentioned that not all clouds yield precipitation but no precipitation is possible without cloud. Clouds play major role in the heat budget of the earth and the atmosphere as they reflect, absorb and diffuse some part of incoming shortwave solar radiation and absorb some part of outgoing longwave terrestrial radiation and then re-radiate it back to the earth's surface.

Classification of Clouds

There is wide range of variations in clouds in terms of height, shape, colour, and transmission or reflection of light. The World Meteorological Organization presented a detailed International Cloud Atlas wherein clouds have been classified into genera (10), species (26), and varieties (31). On an average, all clouds of troposphere are classified into 4 families (high, middle, and low clouds and clouds of vertical development).

*(A) High clouds (height **6** to **20** km)*

(1) cirrus clouds

(2) cirro-cumulus clouds

(3) cirro-stratus clouds

*(B) Middle clouds (height **2.5** to **6** km)*

(4) alto-stratus clouds

(5) alto-cumulus clouds

(6) nimbo-stratus clouds

(C) Low Clouds (height, ground surface to IS km)

(7) strato-cumulus clouds

(8) stratus clouds

(9) cumulus clouds

(10) cumulo-nimbus clouds

(1) *Cirrus Clouds :* The high altitude detached clouds having fibrous (hair like) or silky appearance are called cirrus clouds. They are composed of tiny ice crystals and are transparent and white in colour but have brilliant colours at sunset and sunrise. These clouds are indicative of dry weather.

(2) *Cirro-cumulus clouds* are white coloured clouds having cirriform layer or patches of small white flakes or small globules which are arranged in distinct groups, or wavelike form. They generally appear as ripples similar to sand ripples in the desert. These are not common type of clouds.

(3) *Cirro-stratus clouds* are generally white in colour and spread in the sky like milky thin sheets. In fact, cirro-stratus is a thin white veil of cirrus cloud. They are composed of tiny ice cry stalls which refract the lights of the sun and moon and thus halos are formed around them (around sun and moon). They are so transparent that the sun and moon are visible through them.

(4) *Alto-stratus clouds* are thin sheets of gray or blue colour having fibrous or uniform appearance. When they become thick sheets the sun and moon are obscured and they appear as bright spots behind the clouds. They do not form halos around the sun and moon. They cover the sky partly or totally or are smoothly distributed over the entire sky. They yield widespread continued precipitation either in the form of drizzle or snow.

(5) Alto-cumulus clouds are characterized by wavy layers of globular form. They form fairly regular patterns of lines, groups or waves. In fact, they are individual masses of clouds which are fitted closely together in geometrical patterns. High globular groups of alto-cumulus are some times called as sheep clouds or wool pack clouds. They appear white or gray in colour.

(6) *Nimbo-stratus clouds* are low clouds of dark colour very close to the ground surface. They are so compact and thick (hundreds of metres) that there is complete darkness and there is copious precipitation. Nimbo is from Latin word 'nimbus' meaning thereby rainstorm. They are associated with rain, snow and sleet but are not accompanied by lightning, thunder or hailstorm.

(7) Strato-cumulus clouds are of grey or whitish colour. They are found in rounded patches between the height of 2500m to 3000m. They are composed of globular masses or rolls which are generally arranged in lines, waves or groups. They are generally associated with fair or clear weather but occasional rain or snow is not ruled out.

(8) *Stratus clouds* are dense, lowlying fog-like clouds of dark grey colour but are seldom close to the ground

surface. They are composed of several uniform layers. When these clouds are associated with rains or snow, they are called nimbostratus clouds as referred to above.

(9) *Cumulus clouds* are very dense widespread and dome-shaped and have flat bases. They are white woolpack cloud masses and are associated with fair weather but some times they become thunder clouds.

(10) Cumulo-nimbus clouds are thunder-storm clouds. They show great vertical development and produce heavy rains, snow or hailstorm accompanied by lightning, thunder and gusty winds.

Clouds consist of tiny particles of water or ice, which float in masses at various heights ranging from ground-level (where they occur in the form of fog) to the highest wisps at 12,000 m (40,000 ft).

Measurement and Recording

The amount and nature of the cloud cover is recorded at meteorological stations, and appears on the weather-maps. The amount is recorded in terms of the proportion of the sky which is covered, expressed in eighths (*oktas*), and indicated on the weather-map by a shaded disc. The height and nature of the clouds is shown by means of symbols and figures, according to an international code. If statistics are taken in sufficient detail over a long enough period of time, the mean annual or monthly degree of cloudiness for each station can be plotted, and lines of equal cloudiness (*isonephs*) can be interpolated.

Classification of Clouds

Clouds may be classified either according to their *height* into three groups—high, medium and low cloud, or according to their *general form and appearance* also into three groups—the feathery or fibrous types (*cirrus* or *cirriform*), the globular or heaped types (cumulus or cumuliform), and the sheet or layer types (*stratus* or *stratiform*).

These simple classifications are, however, inadequate to describe the variety of cloud forms. The International Cloud Code lists twenty-eight different types, but ten fundamental *genera* can be recognized. These are distinguished partly by combinations of the three form-names mentioned, partly by adding the suffix *'alto'* to indicate height, and partly by adding *'nimbus'* to signify rain. The genera are subdivided into fourteen *species*, based on shape and structure. These include the lens-shaped *lenticularis*, the turreted *castellanus* and the ragged *fractus*. The genera may be further subdivided into nine varieties, in terms of arrangement and transparency, such as opacus (a thick masking sheet) and translucidus (through which the sun or moon is visible). Nine *supplementary* and *accessory* forms are distinguished, such as arcus (a low dark arch). Leaving these variants aside, the genera are listed in the following Table 2.4:

Table 2.4

Relative position	*Heights of base(m)*	*Name*
High clouds	6000-12,000	Cirrus (Ci) Cirrocumulus (Cc) Cirrostratus (Cs)
Middle clouds	2000-6000	Altocumulus (Ac) Altostratus (As)
Low clouds	Up to 2000,	Stratocumulus (Sc)
	usually less	Stratus (St) Nimbostratus (Ns)
With considerable	–	Cumulus (Cu)
vertical development		Cumulonimbus (Cb)

Description of Cloud Forms

The student is referred to the W.M.O. *International Atlas of Clouds*, which contains illustrations, definitions and descriptions of the ten fundamental genera and their variants.

Cirrus is a delicate fibrous or wispy cloud, consisting of tiny spicules of ice. It is often a fair-weather cloud, though if it is succeeded by cirrostratus it may be an indication of an approaching depression. Where the cirrus is drawn out in a 'mare's tail', it signifies strong winds in the upper atmosphere. *Cirrostratus* is a more complete milky layer of high cloud through which the sun shines with a distinct halo, while *cirrocumulus* forms lines of small globular clouds with a rippled appearance, sometimes called a 'mackerel sky'.

Altostratus is a greyish sheet-cloud, much denser than cirrostratus, but through which it is still possible to see the sun, in the words of the official simile, 'as through ground-glass'. The sky often has a 'watery look', and as this cloud so often heralds the approach of a warm front the appearance is usually not belied. *Altocumulus*, a fleecy cellular cloud in bands of globular patches separated by blue sky, is usually, though not always, a sign of fair weather. If, however, the masses develop into either castellanus or lenticularis, it may mean heavy rain or thunder.

The low varieties of stratus consist of grey uniform sheets of cloud. Stratus itself forms a depressing, heavy grey pall; if continuous rain is falling from it, as at the warm front of a depression, it is termed *nimbostratus*. *Stratocumulus* is a darker, lower and heavier type of altocumulus, which often covers the sky for long dreary periods in winter. Occasionally clearly defined rays of sunshine, known as crepuscular rays, appear to fall to earth through chinks in the cover.

Cumulus is the convection cloud, which grows when rising airmasses reach the level at which condensation takes place, hence the horizontal base of the cloud (Fig. 195). Many of these cumuli develop into large white globular masses, but usually they are fair weather clouds and die away in the evening. If, however, the rising air currents are sufficiently strong, the cumulus cloud continues to grow to an immense vertical height, sometimes upward from a base at about 450m (1500 ft) for 10.11 km (6.7 miles). These towering ,clouds are *cumulonimbus*, and from them heavy showers of rain or hail may fall, accompanied by thunderstorms, hence the name 'thunder-head'. Although from the side the cloud is dazzling white, from below its base may be almost black. The upper parts of a large cumulonimbus cloud. may spread out in the form of an anvil in the direction of the high ˙ vel winds. Ice and snow crystals falling below the spreading layer produce the characteristic wedge-shaped cloud.

The meteorologist pays close attention to the nature of the clouds, for this affords evidence of the trend of the weather, particularly when the sequence of developing cloud forms is studied. For example a definite sequence of cloud is visible during the passage of a depression. Heralded by cirrus and cirrocumulus, a milky pall of cirrostratus then veils the sky, to thicken into altostratus. As the warm front appears, low stratus and nimbostratus from which rain is falling cover the sky. As the cold front arrives, the sharply undercutting cold air may cause cumulonimbus to develop, with heavy showers of rain, to be succeeded by more ragged fractocumulus in a rain-washed hard blue sky as the weather gradually clears.

World Distribution of Cloud

The rainy parts of the world, such as the Equatorial and the Cool Temperate West Marginal types experience a considerable amount of cloud. There is a diurnal cycle of cloudiness in equatorial latitudes—a clear morning sky is slowly obscured by cumulus and cumulonimbus, though by evening the sky begins once again to clear.

The cool mid-latitude zone in the southern hemisphere, the latitudes of the 'Roaring Forties' is one of the cloudiest parts of the world. So, too, are the mountainous areas in the northern hemisphere—the coasts of Norway, western Britain, Ireland and British Columbia. The continental interiors in high latitudes have clear skies in summer, but are subject to long periods of anticyclonic 'gloom', with much low stratus cloud, in winter.

The least cloudy parts of the world are the hot deserts, which are also the most sunny.

In 1960 the U.S. 'Tiros I' satellite produced the first high-altitude photographs of cloud-patterns. In August 1964 'Nimbus', the U.S. weather satellite, was launched into orbit, and its television cameras covered every part of the earth's surface in 24 hours, photographing cloud patterns.

The first complete view of the world's cloud cover was made for 13 February 1965. This was obtained by the U.S. Weather Bureau by assembling 450 individual photographs taken by the cameras carried by another weather satellite, 'Tiros IX', during a period of 24 hours. The various U.S. 'Tiros' and 'Nimbus' and the U.S.S.R. 'Meteor' Satellites have now provided an immense number of cloud photographs. *Nephanalyses* (or cloud-charts) are prepared regularly by the U.S. Weather Bureau.

Measurement and Recording

The amount of rainfall, in units of inches or millimetres, is the theoretical layer of water that would cover the level ground, assuming that none is lost by evaporation, runoff or percolation. One inch of rain is equivalent to 100·9 tons of water per acre, 14,460,000 gallons per sq mile, or 4·2 × 106 litres per sq km; I mm of rain is equivalent to 1 kg per sq m.

A *rain-gauge* consists of a metal cylinder within which a funnel leads into a collecting vessel. This vessel is emptied periodically into a measuring cylinder, graduated so that the depth of rainfall can be read off directly. Should an ordinary measuring cylinder graduated in cubic cm be used, it is necessary to divide this reading by the funnel area in sq cm. The gauge must be very carefully sited; the standard height of the funnel rim is 0.3 m (1 ft) above the ground, and the instrument should be placed well away from trees, high rocks or buildings; it should if possible be twice as far from the nearest building as the latter is high. The gauges are emptied and measured at regular intervals, at an observatory once a day or more frequently. In recent years many gauges have been placed among the British mountains; these are much less accessible and many are visited only monthly.

Self-recording (or *tipping-bucket*) gauges are now used more commonly. The water accumulates in a container which is carefully balanced, and when 5 mm have been collected the water is automatically tipped off. This movement is connected to a pen which traces an ink line on a rotating drum; it shows the rise to 5 mm, then there is a vertical drop, and the rise begins again.

Compilation of Rainfall Records

From rain-gauge measurements all over the world a wide variety of records is compiled of interest to the geographer. For general purposes the *monthly means* of rainfall are either simple arithmetic means of the total rainfall for each month over a period of years, or are weighted figures in which all months are reduced to an equal length to avoid false deductions (as, *e.g.*, if a comparison has to be made between February and August). Monthly means are the basis of the graphs used in and they enable seasonal rainfall to be examined. Annual figures, both means and for actual years, are also useful.

Dispersion diagrams, in which each year's rainfall is shown by placing a dot on a vertical scale, enable one to see at a glance the whole range of wet years, average years and dry years over a long period of time. This is often much more informative than an arithmetic mean, which disguises these differences, of such importance to the people of that area. On 'flood years' and 'famine years' can be distinguished. Another way of indicating these annual variations is in terms of the departure from the annual mean. In India a deficiency of 25 per cent will injure the crops, while one of 40 per cent will cause widespread famine.

The number of rain-days (days with more than 0.25 mm (0.01 in)), and of *wet-days* (days with more than 1.0mm (0.04 in)) are both useful; Hassness (near Buttermere in the English Lake District) has 228 rain-days per annum. *Rain-spells, wet-spells, droughts* and *dryspells*, all defined by standard practice, are recorded in some countries. The longest consecutive run of rain-days in Britain was at Eallbus in the Isle of Islay, which had no less than 80 in 1923. An *absolute drought* is defined as a period of at least fifteen consecutive days each with less than 0.2 mm; the British record was 73 days in London during the spring of 1893. A *partial drought* is a spell of 29 consecutive days, some of which may have slight rain, but during which the total rain does not average more than 0-2 mm per day.

Intensify of rainfall is important, since the rate at which rain falls (ranging from a fine drizzle to a torrential downpour), is related to problems of runoff, percolation into the soil, evaporation, soil erosion and flood control. Information about intensity of rainfall is almost as vital to an understanding of rainfall regimes as are totals. The intensity can be calculated by dividing the total rainfall over a given period by the number of hours during which rain fell (giving the hourly intensity), or by the number of rain-days (giving the daily intensity). The hourly intensity of rain at Boston (U.S.A.) is 0.91 cm (0.36 in), compared with 10.6 cm (4.17 in) at Cherrapunji in north-eastern India. Cherrapunji once experienced 103.6 cm (40.79 in) in a period of 24 hours; *Baguio* in the Philippines has recorded 116.8 cm (46.0 in); and the world record for 24 hours' precipitation was on 16 March 1952, when a station situated at a height of nearly 1200 m (4000 ft) in the island of Reunion in the Indian Ocean had 186 cm (73.6 in). At *Unionville* (Maryland, U.S.A.) on 4 July 1956 3-12 cm (1.23 in) fell in 1 minute.

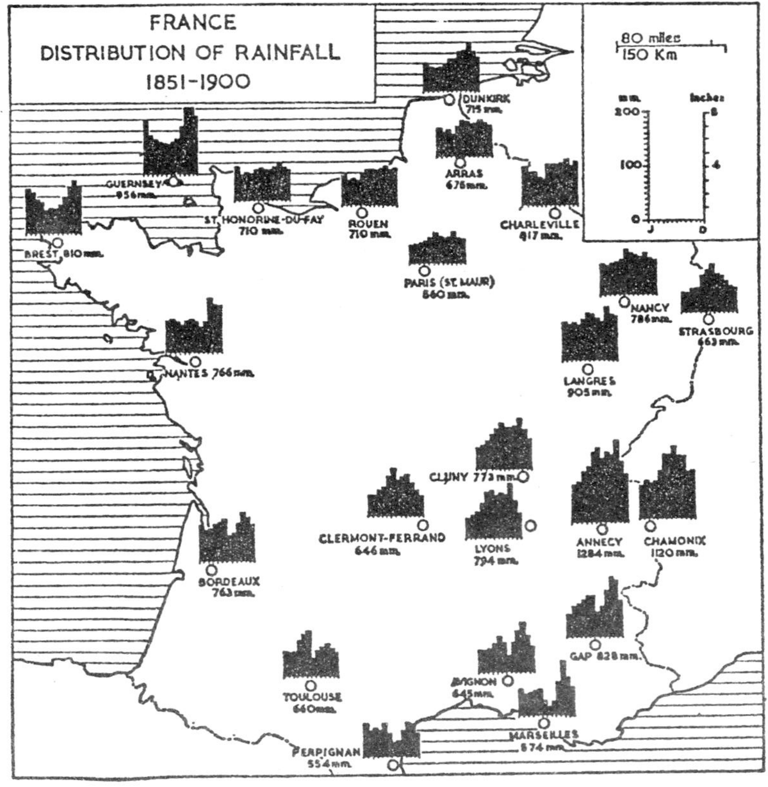

Fig. 2.7 : Rainfall map of France.

Presentation of Rainfall Records

The range of rainfall records may be depicted, graphically and cartographically, in a variety of ways. The most common graphical device to show mean monthly rainfall uses vertical columns, one for each month; they may be shaded or filled black for effect. These columnar diagrams may be strikingly located on a map in the approximate position of the station to which they refer (Fig. 2.7). *Isohyets*—lines on a map to indicate the mean annual, seasonal or monthly rainfall total—are interpolated after values for available stations have been plotted.

The columnar diagrams enable the monthly distribution of rainfall to be seen at a glance. Note the summer drought in the south and the continental summer maximum in the east.

OTHER FORMS OF PRECIPITATION

1. ***Soft Hail :*** It consists small spherical grains formed by the aggregation of tiny ice particles, deposited by direct freezing from water vapour. The hail is white and opaque in character.

   Hail : Hail consists of large pellets or spheres of ice. In fact, hail is a form of solid precipitation wherein small balls or pieces of ice, known as hailstones, having a diameter of 5 to 50mm fall downward known as hailstorms. Hails are very de-structive and dreaded form of solid precipitation because they destroy agricultural crops and claim human and animal lives. True hail is a form of precipitation associated with extreme instability. It falls from lofty cumulonimbus clouds, frequently at the passage of a cold front, and after exceptional local heating and convectional overturning. When droplets form in the lower part of a cloud, they may coalesce with others to form large drops. If the ascending air currents are particularly strong, the drops may be carried upward to a point at which they freeze into ice-pellets. In addition, drops of supercooled water on colliding with tiny ice particles immediately freeze around them as a layer of clear ice. These pellets grow by being carried still higher, so that vapour freezes directly on to them as ice crystals. As the strength of the vertical uprush in a turbulent cloud is extremely variable, the pellets may fall for some distance, partially melt at lower levels, and then be carried upward once again. This may happen several times until the weight of each hailstone is sufficiently great to overcome any up-rising current, and they then fall to earth. This procedure helps to explain the fact that if a large hailstone is cut across, it will probably be found to consist of concentric layers of the clear ice of directly frozen water alternating with the white opaque ice of frozen vapour; as many as twenty-four layers have been counted.

A more recent and indeed more acceptable theory involves the concept of the 'vertical cell' in an atmospheric section within a thundercloud. This postulates that ice-pellets fall out ahead of the updraught, the result of strong upper winds, and are swept up again by the advancing storm before reaching the ground. In other words, the formation of hailstones does not depend on an irregular updraught. Hailstones may attain very large sizes, and there have been some fantastic records claimed. The largest hailstone recorded in Britain had a diameter of 5 cm.

Elsewhere stones of a diameter of 10 cm and of 1 kg in weight have definitely been authenticated, although the unconfirmed claims of hailstones in China 30 cm in diameter and 5 kg in weight seem hard to believe.

Hailstones can do great damage to orchards and glass-houses, and in India and the U.S.A. animals and men are occasionally killed. Hail occurs widely in the world, except in the polar regions where thunderstorms are rare, in the equatorial zone where the air temperature is so high that they melt before reaching the ground, and in the hot deserts. Hailstorms are especially common in spring and early summer in mid-latitudes, such as in the U.S.A. and China, but in Britain they are mainly a winter phenomenon. Other areas of frequent occurrence are northern India and the plateau of South Africa.

2. ***Snow*** : *Snow* is formed when water-vapour condenses at a temperature below freezing-point, passing directly from the gaseous to the solid state and forming minute spicules of ice. These particles unite into crystals which basically are either flat hexagonal plates or prisms; both reveal infinite variations of great beauty in their symmetrical patterns.

   If condensation continues, these crystals unite into snow-flakes and where the lower atmosphere is sumcietitly cool they will reach the ground without melting. For snow to fall there just be both plentiful vapour in the atmosphere and a sufficiently low temperature. Its geographical occurrence and the position of the permanent and winter snow-lines in various latitudes have been discussed. The fall of larger snowflakes from the clouds on the ground surface is called *snowfall.* In fact, snowfall is 'precipitation of white and opaque grains of ice'.

   The snowfall occurs when the freezing level is so close to the ground surface (less than 300m from the surface) that aggregations of ice crystals reach the ground without being melted in a solid form of precipitation as snow. Snow which settles on the ground may be very dry and powdery under low-temperature conditions, as in Antarctica, or it may be wet and compact. In the former case about 30 cm of fresh snow are the equivalent of i cm of rain but in the latter 4 or 6 cm of snow will melt to form i cm of water. For record purposes in Britain, snow is melted and included in the total precipitation figure. A tall cylinder is added to gauges in mountain districts, so that the snow can accumulate and then melt into the container; considerable inaccuracies are inevitable, since often the gauge becomes choked with snow, which then may freeze solid and prevent any further addition. In Canada, however, the actual depth of fresh snow is directly measured with a rule.

3. ***Drizzle:*** The fall of numerous uniform minute droplets of water having diameter of less than 0.5 mm is called drizzle. Drizzles fall continuously from low stratus clouds but the total amount of water received on the ground surface is significantly low.

4. ***Sleet :*** *Sleet* refers to a mixture of snow and rain but in American terminology sleet means falling of small pellets of transparent or translucent ice having a diameter of 5 mm or less. Sleet is an intermediate form of precipitation, though there is some difference between the British and American definitions. The former regards it as a mixture of snow and rain, or of partially melted snow, the latter as raindrops which have frozen and partially melted again.

3

AIR PRESSURE AND ATMOSPHERIC CIRCULATION

INTRODUCTION

Air being a physical substance is an admixture of several gases present in the atmosphere and thus, it has its own weight. Thus, the air exerts pressure through its weight. Air pressure is, thus, defined as total weight of a mass of column of air above per unit area at sea level (unit area being one square inch, one square foot, one square centimetre, one square metre etc.). The atmospheric pressure is maximum at sea level. It exerts the weight of 14.7 pounds on the area of one square inch at sea level or 1034 grams (about one kilogram) per square centimetre. One can imagine as to how much weight of overlying air is being carried by every man daily but he does not feel such enormous weight on his head and shoulders because the air present inside human body exerts equal amount of outward pressure which balances the inward atmospheric pressure. Since the atmospheric pressure decreases with increasing altitudes and therefore the balance between the outward pressure exerted by the air of human body and inward pressure exerted by the atmosphere is disturbed, with the result man suffers from nose and ear bleed at higher altitudes on the mountains.

Air pressure is measured in terms of height of mercury in the glass tube. The standard air pressure at sea level is 1013.25 mb (millibars, millibar is a force equal to 1000 dynes per cm^2 whereas a dyne is a unit offered approximately equal to the weight of one milligram) or 29.92 inches or 76 cm at a temperature of 15°C at the latitude of 45°. The height of mercury upto 0.1 inch in the glass tube is equivalent to 3.4 mb. Air pressure is measured with the help of mercurial barometer (Fortin's barometer), aneroid barometer, altimeter (altitude barometer), barograph, microbarograph etc. The lines joining the places of equal pressure at sea level are called *isobars*. Air pressure decreases with increasing altitudes at the rate of 0.1 inch or 3.4 mb per 600 feet but this rate of decrease is confined to the altitude of a few, thousand feet only. Normally, half of the total atmospheric pressure is confined to the altitude of 1800 feet.

Air pressure also varies seasonally, diurnally and spatially. On an average, it varies from 982 mb to 1033mb. The highest sea level pressure of 1075.2 mb was recorded at Irkutsk in Siberia on 14 January, 1893 while the lowest sea level pressure of 877 mb was recorded in Marina Islands. There is inverse relationship between temperature and pressure *i.e.*, if temperature increases pressure decreases and *vice-versa*. It is very often said that if thermometer is high, barometer is low (pressure is low) and if thermometer is low, barometer is high (pressure is high). The distribution of air pressure is controlled by altitude, temperature, air circulation, rotation of the earth, water vapour etc. The rate of change of pressure per unit horizontal distance is called pressure gradient.

HORIZONTAL DISTRIBUTION OF AIR PRESSURE AND PRESSURE BELTS

The horizontal distribution of air pressure on the globe is studied on the basis of isobars. Air pressure is generally divided in two types viz. :

(1) high pressure, also called as 'high' or anticyclone, and

(2) low pressure, also called as 'low' or cyclone or depression.

If we look at the globe then it appears that there is certain definite system of high and low pressure. If, for generalization, the globe is considered to be homogeneous (either of land or water), then there should be regular and systematic zonal distribution of high and low pressure but the regularity of pressure belts is disturbed due to unequal distribution of land

and water on the globe. The pressure belts are discontinued in the northern hemisphere and several centres of pressure belts are developed but the pressure belts are found more or less in regular pattern in the southern hemisphere.

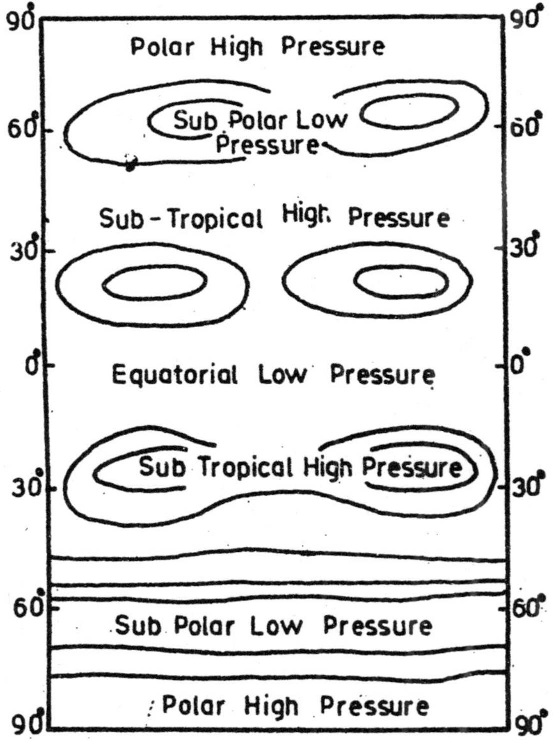

Fig. 3.1: Generalised distribution of air pressure.

There is no definite trend of distribution of pressure from equator towards the poles. If the air pressure would have been the function of air temperature alone there should have been regular increase of pressure poleward because temperature regularly decreases from the equator towards the poles but this is not the case. There is low pressure near the equator due to high mean annual temperature but the existence of high pressure belts near the tropics of Cancer and Capricorn cannot be explained on the basis of temperature because the tropics record very high temperature and hence there should have been low pressure if the temperature would have been the only control of air pressure. The air pressure should increase poleward from the tropics of Cancer and Capricorn because there is rapid rate of decrease of temperature poleward but we find low pressure belt near 60° latitude. Again we find high pressure belts near the poles due to exceedingly low temperature throughout the year. It is obvious that pressure belts are not only induced by thermal factor but they are also induced by dynamic factors.

In an, there are seven pressure belts on the globe. On the basis of mode of genesis pressure belts are divided into two broad categories *e.g.* :

(1) Thermally induced pressure belts (*e.g.* equatorial low pressure belt and polar high pressure belt), and

(2) Dynamically induced pressure belts (*e.g.* subtropical high pressure belt and subpolar low pressure belt (Fig. 3.1).

(1) Equatorial Low Pressure Belt

The equatorial low pressure belt is located on either side of the geographical equator in a zone extending between 5°N and 5°S latitudes but this zone is not stationary because there is seasonal shift of this belt with the northward (summer solstice) and southward (winter solstice) migration of the sun.

During northern summer this belt extends upto 20°N in Africa and to the north of tropic of Cancer in Asia while during southern summer this low pressure belt shifts to 10° to 20°S latitude. The equatorial low pressure belt is thermally induced because the ground surface is intensely heated during the day due to almost vertical sun's rays and thus the lowermost layers of air coming in contact with the heated ground surface also gets wanned.

Thus, wanned air expands, becomes light, and consequently rises upward causing low pressure. The equatorial low pressure belt represents the zone of convergence of north-east and south-east trade winds. There are light, feeble and variable winds within this convergence belt. Because of frequent calm conditions this belt is called a *belt of calm* or *doldrum*.

(2) Sub-Tropical High Pressure Belt

Sub-tropical high pressure belt extends between the latitudes of 25°-35° in both the hemispheres. It is important to note that this high pressure belt is not thermally induced because this zone, besides two to three winter months, receives fairly high temperature throughout the year. Thus, this belt owes its origin to the rotation of the earth and sinking and settling down of winds. It is, thus, apparent that the sub-tropical high pressure belt is dynamically induced. The convergence of winds at higher altitude above this zone results in the subsidence of air from higher altitudes. Thus, descent of winds results in the contraction of their volume and ultimately causes high pressure. This is why this zone is characterized by anticyclonic conditions which cause atmospheric stability and aridity. This is one of the reasons for the presence of hot deserts of the world in the western parts of the continents in a zone extending between 25°-35° in both the hemispheres. This zone of high pressure is called 'horse latitude' because of prevalence of frequent calms. In ancient times, the merchants carrying horses in their ships, had to throw out some of the horses while passing through this zone of calm in order to lighten their ships. This is why this zone is called horse latitude. It is interesting to note that this zone of high pressure is not a continuous belt but is broken into a number of high pressure centres or cells (Fig. 3.1)

(3) Sub-Polar Low Pressure Belt

This belt of sub-polar low pressure is located between 60°-65° latitudes in both the hemispheres. The low pressure belt does not appear to be thermally induced because there is low temperature throughout the year and as such there should have been high pressure belt instead of low pressure belt. It is, thus, obvious that this low pressure belt is dynamically produced. In fact, the surface air spreads outward from this zone due to rotation of the earth and low pressure is caused. It may be pointed out that this factor should be more effective at the poles but the effects of the rotation is negated or say overshadowed due to exceptionally low temperature prevailing throughout the year at the poles. The sub-polar low pressure belt is more developed and regular in the southern hemisphere while it is broken in the northern hemisphere (Fig. 3.1) because of overdominance of water (oceans) in the former. Instead of regular and continuous belt there are well defined low pressure centres or cells over the oceans in the northern hemisphere *e.g.* in the neighbourhood of Aleutian Islands in the Pacific Ocean and between Greenland and Iceland in the Atlantic Ocean. It may be noted that due to great contrasts of temperatures of the continents and oceans during northern summer the low pressure belt becomes discontinuous and is found in a few low pressure cells while the temperature contrast between the continents and oceans is much reduced during winter and hence low pressure belt becomes more or less regular and continuous in the northern hemisphere.

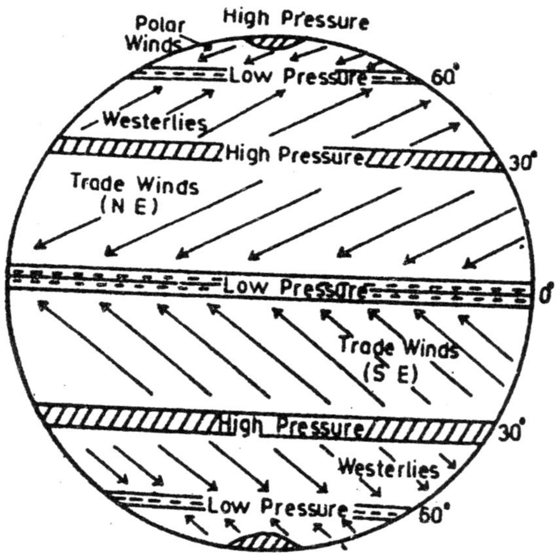

Fig. 3.2: Air pressure and wind belts.

(4) Polar High Pressure Belt

High pressure persists at the poles throughout the year because of prevalence of very low temperature (below freezing point) all the year round. In fact, both the factors, thermal and dynamic, operate at the poles. There is thinning out of layers of air due to diurnal rotation of the earth as the air spreads outward due to this factor but this effect is overshadowed by thermal factor and hence high pressure is produced due to very low temperature.

The horizontal distribution of air pressure is represented and studied through isobars for the months of July (to represent pressure conditions during summer season) and January (to represent air pressure during winter season in the northern hemisphere). Figs, 3.3 and 3.4 portray the world distribution of air pressure through isobars in July and January respectively. The class interval of isobars is 3mb.

SHIFTING OF PRESSURE BELTS

The surface pattern of air pressure as discussed above and represented through Figs. 3.1 and 35.2, seldom remains stationary in its latitudinal zone. There are daily, seasonal and annual changes in air pressure because of northward and southward movement of the overhead sun (summer and winter solstices), contrasting nature of the heating and cooling of land and water etc. The lowest pressure is developed between 2 to 4 P.M. during the day due to maximum temperature while highest pressure is recorded between 4-6 A.M. due to minimum temperature during night. Coastal land records low pressure while adjoining oceanic area has high pressure during the day. This situation is reversed during night. Except polar high pressure belt all the pressure belts move northward with the northward movement of the sun during summer solstice. On the other hand, except the polar high pressure belt, all the belts move southward due to southward movement of the sun during winter solstice when the sun is vertical at the tropic of Capricorn. The pressure belts occupy their normal ideal position at the time of vernal equinox (21 March) and autumnal equinox (23 September) when the sun is vertical at the equator.

WORLD DISTRIBUTION OF PRESSURE

The Planetary System

The theoretical concept of pressure distribution on a homogeneous globe at sea-level (an 'ideal' or 'planetary' system) is useful as an elementary basis, but it must be realized that this is subject to immense modifications. Within a few degrees of the Equator is a belt of pressure generally less than 1013 mb (29.9 in), the Equatorial Trough or low-pressure belt, now referred to as the *Intertropical Convergence Zone*. This is an area of high temperature and high humidity, commonly known as the Doldrums, where the air near sea-level is stagnant or sluggish.

The low pressure is due to heating, for the pressure of a volume of air decreases as its temperature rises. Increasing numbers of readings from high-altitude observations show that these sluggish conditions rapidly disappear with increasing height, and air-stream flows of considerable velocity can be discerned. It must be emphasized that this is a generalization on a planetary scale, and there are considerable regional modifications.

At about latitudes 30°N and S occur the Subtropical high-pressure belts, sometimes known as the 'Horse Latitudes', zones of calms and of descending air currents. It is not easy to explain the existence of these belts; they may in part be due to a general poleward movement from the Equator of air in the upper part of the troposphere, which comes under the influence of the earth's rotation so that its centrifugal force throws it back, causing an accumulation in these latitudes. There seems also to be an equatorward movement of air from high latitudes in the upper part of the troposphere, which descends in the Horse Latitudes, and so increases the air accumulation.

Nearer the Poles occur the Subpolar low-pressure belts. One reason for this pressure distribution is that the rotation of the earth causes a polar whirl and therefore a tendency towards low pressure at the Poles. But the intense cold around the Poles causes the thermal effect to overcome the dynamical one, with the result that the low-pressure belts tend to be on and just outside the polar circles.

The Antarctic ice-cap, above which lies intensely cold dense air, has therefore a shallow Polar high-pressure system over it. The Arctic region, however, is more complicated, for it is an ocean basin, with land-masses grouped round two-thirds of its perimeter; low pressure over the North Pole is not uncommon, even on mean monthly charts. the planetary pressure systems at the Equinoxes, but the various pressure belts are displaced some degrees north and south with the seasons. Owing to the apparent moment of the overhead sun through 47° of latitude, the Thermal Equator also moves, but not so much (through about 10°), and the Equatorial low-pressure belt moves with it through about 7°. The other pressure belts more or less move likewise.

Pressure 'cells'

The main modification to this simple planetary concept is imposed by the irregular distribution of land and water, particularly in the northern hemisphere, which causes marked seasonal temperature changes. The result is the interruption of the latitudinal 'belts' by the creation of high pressure over the cold continental interiors in winter and, conversely, of low pressure over the heated continents in summer (Figs. 8.2, 8.3). This effect is not so apparent in the southern hemisphere, since the areas of land are small; the Subtropical high-pressure belt is not markedly interrupted, except in summer by slight southward extensions of the Equatorial low over the southern continents, while the Subpolar low-pressure belt in the Southern Ocean is continuous throughout the year. In the northern hemisphere the pressure belts are so interrupted that they form a series of great 'cells', which may be thought of as large-scale eddies in the atmosphere, often of considerable depth and extending up to the tropopause.

In winter the intense Eurasiatic (Siberian) and American highs dominate the continents, and the North Atlantic and North Pacific lows (sometimes called the Icelandic and Aleutian lows respectively) are over the northern parts of these oceans. In summer, however, lows form over Asia and North America as a result of the heating of the land; the centre of the Asiatic low is in north-western India, of the American low in the south-west of the U.S.A. and northern Mexico. Each is really a northerly extension of the Equatorial low.

These lows interrupt the Subtropical high-pressure belt, but the North Atlantic and North Pacific are dominated by the high-pressure cells of the Horse Latitudes, sometimes called the *Azorean* and *Hawaiian highs.*

The Icelandic and Aleutian lows are less intense, and are situated farther north than in winter; the Aleutian, in fact, virtually vanishes.

The main 'cells' of relatively high pressure and low pressure near sea-level are indicated diagrammatically.

Upper Air Pressure

In recent years there has been much research into conditions in the upper air, and pressure maps showing these conditions are constructed and published as a routine procedure. However, the question of upper-air pressure is extremely complicated, although it is increasingly realized that developments in meteorology (particularly in long-range weather forecasting) are largely connected with the upper air.

A comparison of a series of high-altitude pressure-maps with the corresponding ones for sea-level shows distinct differences in pressure distribution, sometimes actual reversals of conditions, though on the whole the patterns are less complicated than at sealevel, because of the reduced influence of the distribution of land and sea.

Over a large anticyclone at sea-level there may be a low at 3000 m (10,000 ft). At no great height in the middle troposphere (about 10 km, 6 miles), the pressure distribution appears to be dominated by a vast low-pressure system in each hemisphere, approximately centred over each Pole, with a series of high-pressure cells more or less over latitudes 15° N and S.

PRESSURE GRADIENT AND AIR CIRCULATION

The difference of pressure between any two places is called *pressure gradient* Steep pressure gradient is represented by closely spaced isobars while widely spaced isobars reveal low pressure gradient. The direction of pressure gradient is considered from high pressure to decreasing pressure and the pressure direction is always perpendicular to isobars. Pressure gradient is also called *barometric slope.* There is close relationship between pressure gradient and air circulation.

The air moves from high pressure to low pressure. In other words, air movement follows barometric slope. The direction of air movement should be perpendicular to the isobars (Fig. 3.5) because the direction of pressure gradient is perpendicular to the isobars but the direction is deviated from the expected theoretical direction due to coriolis force caused by the rotation of the earth and thus the winds cross the isobars at acute angle instead of right angle.

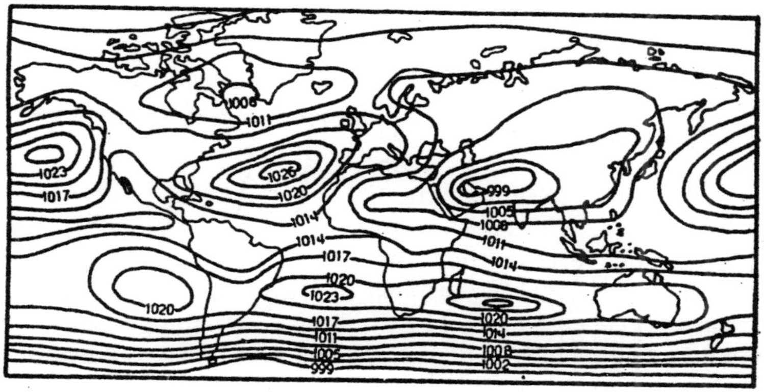

Fig. 3.3 : Isobars and world distribution of air pressure in July, figures in millibars.

Fig. 3.4 : Isobars and world distribution of air pressure in January. Figures in millibars.

Horizontal pressure (pressure gradient), rotation of the earth and coriolis force, frictional force, centrifugal action of wind etc. affect and control air motion. The wind blowing parallel to the isobars generally at the height of 600m is called geostrophic wind.

WIND DIRECTION AND RELATED LAWS

The direction of surface winds is usually controlled by the pressure gradient and rotation of the earth. Because of rotation of the earth along its axis the winds are deflected. The force which deflects the direction of winds is called dêflection force. This force is also called coriolis force on the basis of famous scientist G.G. Coriolis. Because of coriolis force all the winds are deflected to the right in the northern hemisphere while they are deflected to the left in the southern hemisphere with respect to the rotating earth. This is why winds blow counter-clockwise around the centre of low pressure (to make cyclonic circulation) in the northern hemisphere while they blow clockwise in the southern hemisphere.

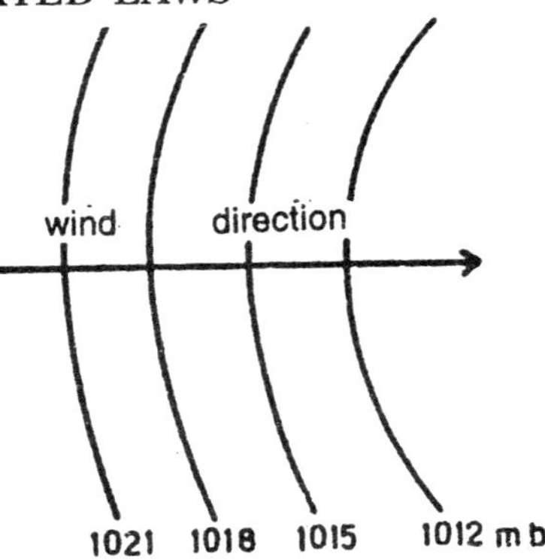

Fig. 3.5 : Pressure gradient and wind direction.

It may be remembered that the direction of pressure gradient is always from high pressure to low pressure. The earth rotates from west to east. Every latitude is a complete circle. Equatorial latitudinal circle is the largest one and the latitudinal circles decrease poleward wherein polar circle is the smallest one. The whole earth completes one rotation along its axis roughly in 24 hours. Thus, the rotational speed of the earth is highest at the equator and decreases poleward. When the wind moves either northward or southward following straight path in equatorial region it does not reach its destination because by that time the destination place moves ahead and the wind lags behind because of high rotational speed of the earth (Fig. 3.6). Contrary to this the wind moving either northward or southward in high latitudes reaches ahead of its destination because of decreasing rotational speed of the earth. Based on the deflective force (coriolis force) of the earth and deflection of winds Ferrel has propounded his law which is popularly known as ***Ferrel's Law***. *This law states that 'if one stands with one's back towards the direction from where winds are coming they (winds) are deflected to the right in the northern hemisphere and to the left in the southern hemisphere.' Buys Ballot, a Dutch meteorologist, formulated his law of wind direction in 1857 on the basis of relationship between wind direction and pressure. According to his law 7n the northern hemisphere if you stand with your back to the wind, there will be low pressure to your left and high pressure to your right. In the southern hemisphere the coriolis deflection being to the left the situation is reversed.'*

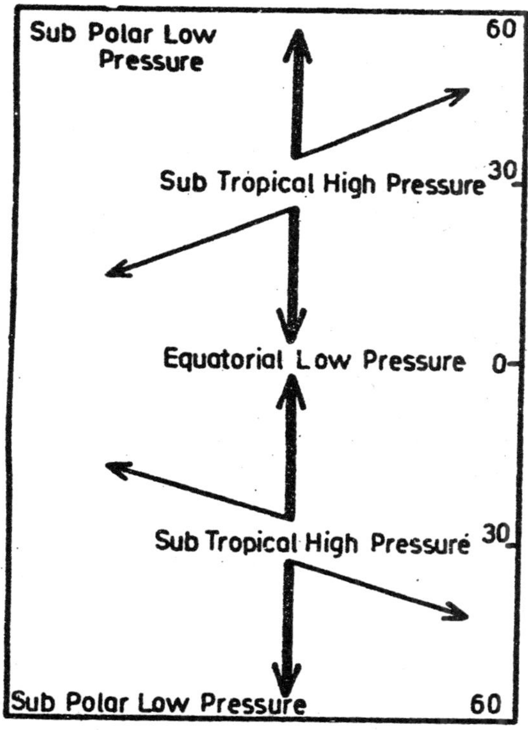

Fig. 35.6: Deflective force and wind direction.

Frictional force not only retards the speed of winds but also deflects them. The frictional force is effective upto a height of a few thousand metres only. The hillslope or ground slope facing the winds is called windward slope or onward side while the opposite slope is called leeward slope.

CLASSIFICATION OF WINDS

The winds blowing almost in the same direction throughout the year are called *prevailing* or *permanent winds*. These are also called as *invariable* or *planetary winds* because they involve larger areas of the globe. On the other hand, winds with seasonal changes in their directions are called seasonal winds (*e.g.* monsoon winds).

Winds blowing in a particular locality are called local winds (*e.g.* chinook, sirocco, harmattan, mistral, bora, blizzard, loo etc.). Winds blowing from hill tops to the valleys and from valley floor to the hill tops are called *mountain* and *valley breezes*. Winds blowing from land to sea and from sea to land are called *land* and *sea breezes*. Thus, the winds are classified into two broad categories *e.g.* :

(1) *permanent winds or invariable winds or prevailing winds* (*e.g.* trade winds, westerlies and polar winds),

(2) and *variable winds*, which are further divided into

   (i) seasonal winds,

   (ii) local winds,

   (iii) mountain and valley breezes,

   (iv) land and sea breezes etc.

Alternatively, atmospheric motion or wind movement is divided into three categories *e.g.* :

(1) Primary circulation including planetary wind systems which are related to global pressure belts (*e.g.* trade winds, westerlies and polar winds),

(2) Secondary circulation consisting of cyclones, anticyclones, monsoons, and air masses, and

(3) Tertiary circulation which includes local winds as referred to above.

Table 3.1 : The Beaufort scale for winds

Beaufort	*Knots*	*Explanatory Terms*	*Observable Description*
0	less than 1	calm	Calm, smoke rises vertically.
1	1-3	Light air	Direction of wind shown by smoke drift but not wind vanes.
2	4-6	Light breeze	Wind [illegible] on face, leaves rustle; ordinary vane move by wind.
3	7-10	Gentle breeze	Leaves and small twigs in constant motion; wind extends light flag.
4	11-16	Moderate breeze	Raises dust and loose paper; small branches are moved
5	17-21	Fresh breeze	Small trees in leaf bigin to sway; wavelet form on inland water.
6	22-27	Strong breeze	Large branches in motions; whistling heard in telegraph wires; umbrellas used with difficulty.
7	28-33	Moderate gale	Whole trees in motion; inconvenience felt when walking against wind.
8	34-40	Fresh gale	Breaks twigs from trees.
9	41-47	Strong gale	Slight structural damage occurs, chimneys and slates carried away.
10	48-55	Whole gale	Seldome experienced inland; trees uprooted, considerable damage to structures.
11	56-63	Storm	Very rarely experienced; accompanied by widespread damage.
12	64-71	Hurricane	Violence and destruction.

PERMANENT OR PLANETARY WINDS

On an averages, the location of high and low pressure belts is considered to be stationary on the globe (though they are seldom stationary). Consequently, winds blow from high pressure belts to low pressure belts. The direction of such winds remains more or less the same throughout the year though their areas change seasonally. Thus, such winds are called permanent winds. Since these winds are distributed all over the globe and these are related to thermally and dynamically induced pressure belts and rotation of the earth and hence they are called *planetary winds*. These winds include trade winds, westerlies and polar winds (Fig. 3.7).

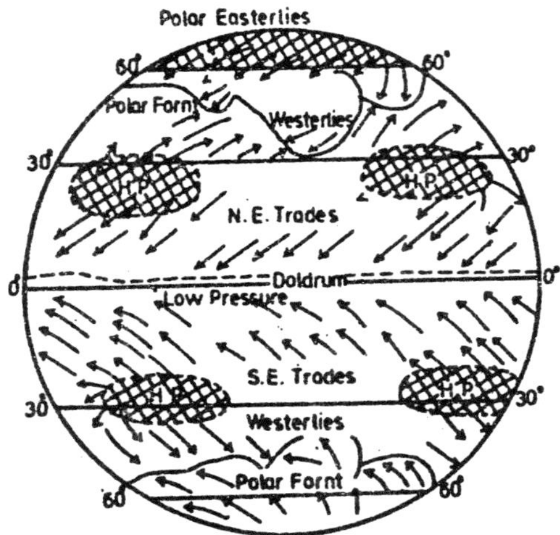

Fig. 3.7: The generalized global pattern of planetary winds.

(1) Winds in the Tropics

Generally, the areas extending between 30°N and 30°S latitudes are included in tropical zone.

Formerly, it was believed that trade winds blow from the subtropical high pressure belts to the equatorial low pressure belt. The north-east and south-east trades converge along the equator and there are upper air anti-trades blowing in the opposite directions of the surface trade winds. The weather conditions throughout the tropical zone remain more or less uniform. There is a belt of calm or doldrum characterized by feeble air circulation. These views of weather conditions prevailing in the tropics are now considered as old concepts and have now been refuted on the basis of new information based on numerous observations made in the upper air and near the earth's surface during and after second world war. Nev discoveries have now put question marks against the views of zonal character and regularity of surface trade winds, uniformity of weather conditions in the tropics, and upper air antitrade winds. It has been discovered that trade winds blow with regularity only over some parts of the tropical oceans (mainly over the eastern parts). Upper air anittrades, contrary to earlier beliefs, are not found every where but are confined to certain areas only. The weather conditions in the tropics are not calm and uniform but they are frequently interrupted by atmospheric disturbances (cyclones, hurricanes, typhoons, sea waves etc.). Thus, the tropical zone is characterized by doldrum, equatorial westerlies, and trade winds.

(i) *Doldrum and equatorial westerlies* : A belt of low pressure, popularly known as equatorial trough of low pressure, extends along the equator within a zone of 5°N and 5°S latitudes. This belt is called the belt of calm or doldrum because of light and variable winds. It was believed that doldrum is a regular feature all along the equator and is characterized by strong convective instability leading to the formation of cumulonimbus clouds and copious rainfall daily but the recent observations have now shown that the belt of doldrum is not continuous but is confined to certain localities only. This belt is subjected to seasonal and spatial variations due to northward and southward movement of the over-head sun (summer and winter solstices). In fact, the belt of doldrum shifts northward during summer solstice (when the sun is vertical (Fig. 3.8) viz. :

(1) Over the tropic of Cancer, 21 June) and comes back to its normal position at the time of winter solstice (when the sun is vertical over the tropic of Capricorn, 23 December). According to *Flohn* the doldrums extend upto 200 longitudes in discontinuous manner. Crowe has identified 3 zones of doldrums *e.g.* Indo-Pacific Doldrum extending from the eastern coast of Africa to 180° longitude for a distance of 16,000 km and covering an area of 25,800,000 km^2, thus, covers about one third of the total length of the equator,

(2) Equatorial Western Coastal Region of Africa, and

(3) Western Coastal Margin of Central America.

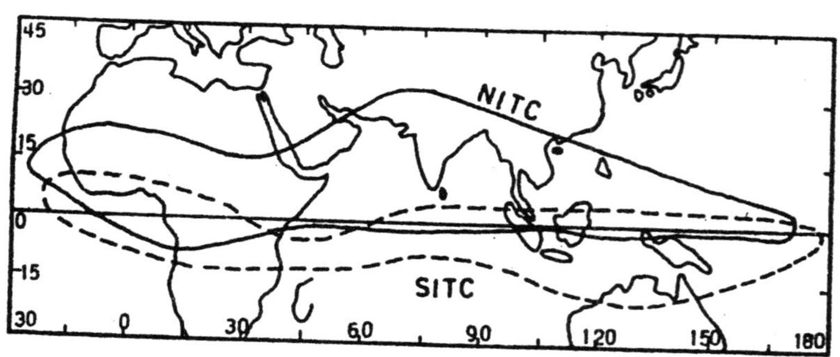

Fig. 3.8 : Intertropical convergence (NITC and SITC).

A few meteorologists and climatologists believe in the existence of cyclonic fronts in the equatorial regions while some refute this concept on the ground that fronts are formed only when two air masses of contrasting temperatures converge but such conditions are not found in the equatorial region. The equatorial or tropical fronts are called *intertropical fronts* (ITF) or *intertropical convergence* (ITC). These fronts represent the meeting ground of north-east and south-east trade winds. The northern and southern boundaries of intertropical convergence are called north intertropical convergence (NITC) and south intertropical convergence (SITC) respectively (Fig. 3.10). There is seasonal shifting in the NITC and SITC with the northward (summer solstice) and southward migration (winter solstice) of the sun.

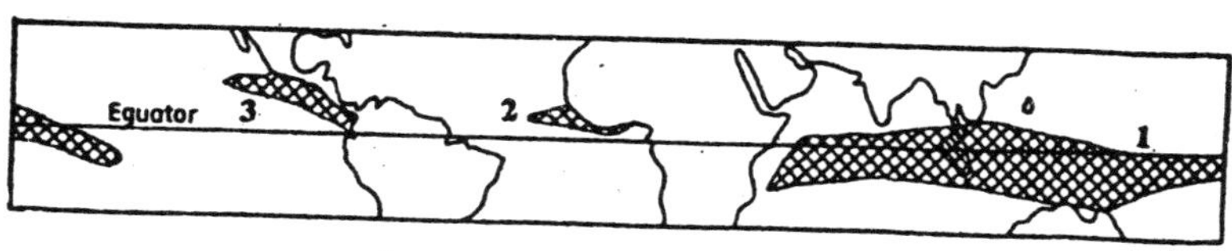

Fig. 35.9 : Position of doldrums- (1) Indo-Pacific doldrum, (2) equatorial western coastal region of Africa, and (3) western coastal region of central America.

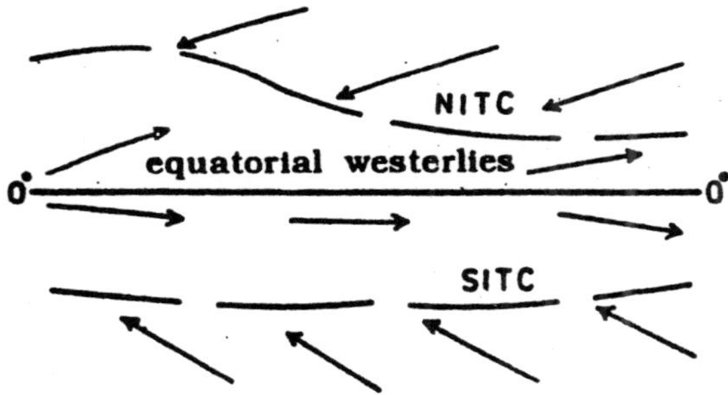

Fig. 3.10: NITC and SITC and Equatorial Westerlies.

On an average, there is westerly air circulation (form west to east) in the doldrums or say in the intertroppical convergence. These westerly winds have been called by *Flohn as equatorial* westerlies (Fig. 3.10) which cover 200° longitudes. According to Flohn the equatorial westerlies cover the areas extending from the western parts of Africa across the Indian Ocean to the western Pacific Ocean. The equatorial westerlies are associated with strong atmospheric disturbances (cyclonic storms). Flohn has further maintained that south-western monsoons are, intact, equatorial westerlies because these winds are extended upto 30-35°N latitudes over Indian subcontinent due to northward shifting of NITC at the time of summer solstice (Fig. 3.9).

(ii) *Trade winds* : There is more or less regular inflow of winds from subtropical high pressure belts to equatorial low pressure belt. These tropical winds have north-easterly direction in the northern hemisphere while they are south-easterly in the southern hemisphere. These winds are called trade winds because of the fact that they helped the sea merchants in sailing their ships as their (of trade winds) direction remains more or less constant and regular. According to Ferrel's law (based on coriolis force generated by the rotation of the earth) trade winds are deflected to the right in the northern hemisphere and to the left in the southern hemisphere (Fig. 3.6).

There are much variations in the weather conditions in the different parts of trade winds.

The poleward parts of the trade winds or eastern sides of the subtropical anticyclones are dry because of strong subsidence of air currents from above. Because of the dominance of anticyclonic conditions there is strong atmospheric stability, strong inversion of temperature and clear sky. On the other hand, the equatorward parts of the trade winds are humid because they are characterized by atmospheric instability and much precipitation as the trade winds while blowing over the oceans pick up moisture. It may be stated that the trade winds are more regular and constant over the oceans than over the lands. At some places on the lands (*e.g.* S.E. Asia and southern USA) the trade winds disappear during summer season due to formation of low pressure cells because of high temperature but the trade winds are more constant and regular over the continents during winter season. It may be pointed out that the zone of trade winds is called *Hadely Cell* on the basis of the convective model prepared by Hadley for the entire earth.

(2) Horse Latitudes And Westerlies

(i) *Horse latitudes :* The dynamically induced (due to subsidence of air currents) subtropical high pressure belt extends between 30°-350 (25°-35°) latitudes in both the hemispheres. Thus, this belt separates two wind systems viz. trade winds and westerlies. It is also apparent that the subtropical high pressure belt is the source for the origin of trade winds (blowing towards equatorial low pressure belt) and westerlies (blowing towards subpolar low pressure belt) because winds always blow from high pressure to low pressure. The air after being heated near the equator ascends and after blowing in opposite direction to the surface trade winds descends in the latitudinal zone of 30°-35°. Thus, the descent of winds from above causes high pressure on the surface which in turn causes anticyclonic conditions. This is why the anticyclonic conditions cause atmospheric stability, dry condition and very weak air circulation. Thus, this zone (30°-35°) is characterized by weak and variable winds and calm. This belt of calm is very popularly known as horse latitudes because of the fact that in ancient times the merchants had to throw away some of the horses being carried in the ships in order to lessen the weight so that the ships could be sailed through the calm conditions of these latitudes.

Anticyclones are produced due to subsidence of air currents in the horse latitudes. These anticyclones are known as 'subtropical highs' or subtropical anticyclones, the eastern and western parts of which are characterized by contrasting weather conditions. The eastern parts (spreading over the western parts of the continents) are marked by descent of air currents, inversion of temperature and consequent atmospheric stability and dry conditions. This is why hot and dry tropical deserts are found in the western parts of the continents within the latitudinal zones of 20°-30° in both the hemispheres (*e.g.* Sahara and Kalahari in Africa, Chile-Peru desert or Acatama in South America, Arabian and Thar deserts in Asia, deserts of S.W. USA, and Australian deserts). The western parts of subtropical anticyclones (covering the eastern parts of the continents and western parts of the oceans) are humid because some sort of atmospheric instability is caused due to weakening of air descent (*e.g.* in the areas of Caribbean Sea, Mexican Gulf and adjoining areas, eastern China, southern Japan, south-east Brazil and eastern Australia).

(ii) *Westerlies* : The permanent winds blowing from the subtropical high pressure belts (30°-35°) to the subpolar low pressure belts (60°-65°) in both the hemispheres are called westerlies (Fig. 3.7). The general direction of the westerlies is S.W. to N.E. in the northern hemisphere and N.W. to S.E in the southern hemisphere. There is much variation in the weather conditions in their poleward parts where there is convergence of cold and denser polar winds and warms and lighter westerlies. In fact, a cyclonic front, called as polar front, is formed due to two contrasting air masses as referred to above and thus temperate cyclones are originated. These cyclones move along with the westerlies in easterly direction. Thus, the general characteristic features of the westerlies are largely modified due to cyclones and anticyclones associated with them. Because of the dominance of land in the northern hemisphere the westerlies become more complex and complicated and become less effective during summer seasons and more vigorous during winter season. These westerlies bring much precipitation in the western parts of the continents (*e.g.* north-west European coasts) because they pick up much moisture while passing over the vast stretches of the oceans. The westerlies become more vigorous in the southern hemisphere because of lack of land and dominance of oceans. Their velocity increases southward and they become stormy. They are also associated with biosterous gales. The velocity of the westerlies becomes so great that they are called roaring forties between the latitudes of 40°-50°S, furious fifties at 50°S latitude and shrieking sixties at 60°S latitude.

(3) Polar Winds

A low pressure belt, produced due to dynamic factor, lies within the latitudinal belt of 60°-65° in both the hemispheres. This belt of low pressure is more persistent in summer season but generally disappears in winter season. The Icelandic and Aleutian low pressure cells persist throughout the year. There is very high pressure over the poles because of exceedingly low temperature. Thus, winds blow from polar high pressure to subpolar low pressure cells. These are called polar winds which are north-easterly in the southern hemisphere and south-easterly in the southern hemisphere. The zone of polar winds shrinks due to northward shifting of pressure belts at the time of northern summer (summer solstice) in the northern hemisphere but it is extended upto 60°N latitude during northern winter (winter solstice).

SEASONAL SHIFTING OF WIND BELTS AND THEIR CLIMATIC SIGNIFICANCE

In the absence of the revolution of the earth around the sun in about 365 days the global pressure belts would have been permanent and stationary at their places but the relative position of the earth with the sun changes within a year due to earth's revolution and thus the position of all the pressure belts except the polar high pressure belts changes with the northward and southward migration of the sun. At the time of summer solstice the sun is vertical over the tropic of Cancer (June 21) and therefore all the pressure belts except the northern polar high pressure belt shift northward (fig. 3.11). The equatorial low pressure belt prevails between 0° latitude (equator) and 10°N latitude, subtropical high pressure belt extends between 30°-40°N latitudes. Thus, all the wind belts associated with the said pressure belts also shift northward. The sun becomes vertical over the equator at the time of autumnal equinox (23 September) and hence all the pressure belts which shifted to the north occupy their normal positions (Fig. 3.11). After this there is southward migration of the sun which becomes vertical over the tropic of Capricorn at the time of winter solstice (23 December) and hence the pressure and wind belts shift southward except the southern polar high pressure belt. Thereafter the sun again becomes vertical over the equator at the time of vernal equinox (21 March) and hence all the pressure and wind belts occupy their normal positions (Fig. 3.11), thus, there is shifting in the positions of the pressure and wind belts due to seasonal changes of position of the earth in relation to the sun. These seasonal changes in the relative positions of the pressure and wind belts introduce the following typical climatic conditions.

(i) The Mediterranean climatic regions are found in the western parts of the continents within the latitudinal zone of 30°-45° in both the hemispheres. The subtropical high pressure belts extending between 30°-35° latitudes are characterized by dry trade winds during summer season because mostly they come from over the land. More over, this zone is characterized by anticyclonic conditions due to descent of air currents. This belt extends upto 40° latitudes in the northern hemisphere at the time of summer solstice and in the southern hemisphere at the time of winter solstice. Thus, the western parts of the continents within the zone of 30°-40° latitudes do not receive rainfall during summer season. On the other hand, the subtropical belt shifts towards the equator at the time of winter solstice in the northern hemisphere and at the time of summer solstice in the southern hemisphere, consequently, the zone between 30°-40° latitude is characterized by westerlies which give much precipitation during winter season because they come from over the oceans. This is why the Mediterranean regions are characterized by dry summers and wet winters and a typical Mediterranean type of climate is produced.

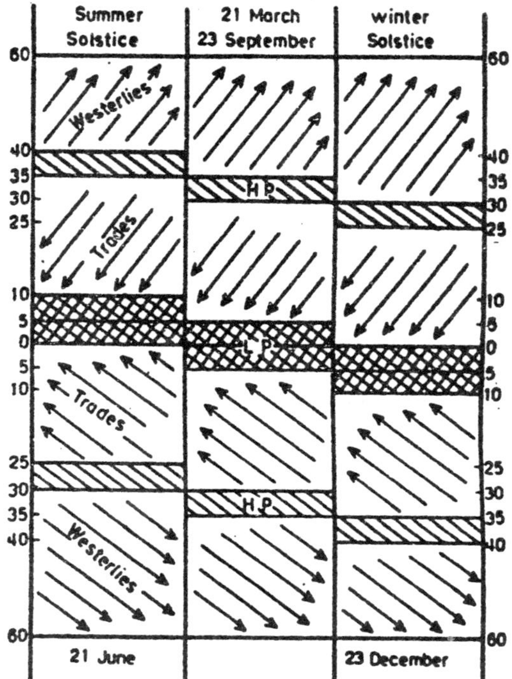

Fig. 3.11 : Shifting of pressure and wind belts.

(ii) The regions lying between 60°-70° latitudes are characterized by two types of winds in a year because of shifting of pressure and wind belts. With the northward migration of the sun at the time of summer solstice the polar easterlies are weakened during northern summer because the westerlies extend over these areas due to northward (poleward) shifting of subpolar low pressure belt while the situation is quite opposite in the southern hemisphere because the polar easterlies extend over much of the areas of the westerlies due to equatorward shifting of subpolar low pressure belt. The situation is reversed at the time of winter solstice when there is southward migration of the sun. Thus, the polar easterlies are re-established over the areas lying between 60°-70° latitudes in the northern hemisphere because of the shifting of the belt of westerlies southward. Consequently, a typical climate characterized by wet summers through westerlies and associated cyclones and dry winters due to polar castelies is produced.

(iii) Monsoon climate is the result of the shifting of pressure and wind belts. Due to northward migration of the sun in the northern hemisphere at the time of summer solstice the north intertropical convergence (NITC) is extended upto 30°N latitude over Indian subcontinent, south-east Asia and parts of Africa. Thus, the equatorial westerlies are also extended over the aforesaid regions. These equatorial westerlies, in fact, become the south-west or summer monsoons. These south-west monsoon winds bring much rains because they come from over the ocean and are associated with tropical atmospheric storms (cyclones). The NITC is withdrawn from over the Indian subcontinent and south-east Asia because of southward shifting of pressure and wind belts due to southward migration of the sun at the time of winter solstice. Thus, north-east trades are re-established over the aforesaid areas. These north-east trades, in fact, are north-east or winter monsoons. Since they come from over the lands, and hence they are dry.

TRICELLULAR MERIDIONAL CIRCULATION OF THE ATMOSPHERE

According to the old concept of the mechanism of general circulation of the atmosphere the movement of air is temperature dependent. In other words, temperature gradient causes air circulation on the earth's surface. According to the advocates of thermal school of the mechanism of general circulation of the atmosphere the tropical areas receive maximum amount of solar energy which substantially decreases poleward. Thus, there is latitudinal imbalance of solar radiation from lower to higher latitudes. Consequently, there is transfer of heat through horizontal air circulation from the areas of high solar radiation (low latitudes) to the areas of low solar radiation (high latitudes) in order to balance the heat energy so that there does not exist too much heat energy in the low latitudes and too low heat energy in the high latitudes. This old school considers only the horizontal component of the atmospheric circulation and does not consider the potential energy generated by unequal heating of the earth and its atmosphere and its continuous transformation into kinetic energy. It may be pointed out that the potential heat energy is continuously transformed into kinetic energy by the upward movement (ascent) and downward movement (descent) of heated and cold air respectively. It may be remembered that the kinetic energy is also dissipated due to friction and small-scale atmospheric disturbances upward. Thus, it is necessary that there must exist balance between the rate of generation of kinetic energy and the rate of its dissipation due to friction. The modern concept of the mechanism of general circulation of the atmosphere, thus, includes both, the horizontal and vertical components of atmospheric circulation.

The modem school envisages a three-cell model of meridional circulation of the atmosphere, popularly known as tricellular meridional circulation of the atmosphere, wherein it is believed that there is cellular circulation of air at each meridian (longitude). Surface winds blow from high pressure areas to low pressure areas but in the upper atmosphere the general direction of air circulation is opposite to the direction of surface winds. Thus, each meridian has three cells of air circulation in the northern hemisphere *e.g.* :

(1) Tropical cell or Hadley cell,

(2) Polar front cell or midlatitude cell or Ferrel cell, and

(3) polar or subpolar cell (Fig. 3.12).

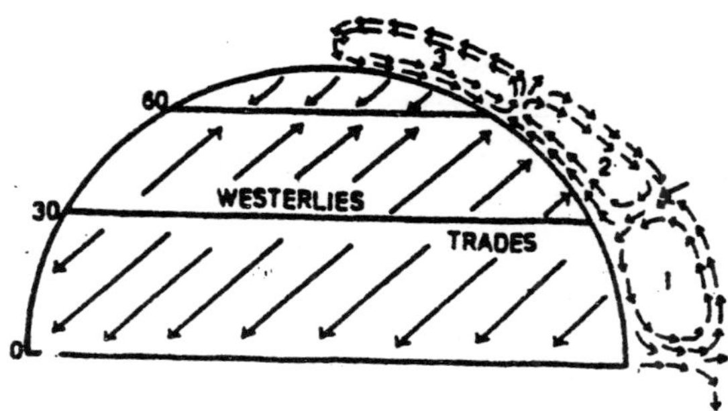

Fig. 3.12 : Tricellular meridional circulation of the atmosphere (1) tropical Hadley cell, (2) Midlatitude cell, and (3) Polar cell.

(1) *Tropical cell* is also called as Hadley cell because G. Hadley first identified this thermally induced cell in both the hemispheres in the year 1735. The winds after being heated due to very high temperature at the equator ascend upward. These ascending warm and moist winds release latent heat after condensation which causes further ascent of the winds which after reaching the height of 8 to 12 kilometers in the troposphere over the equator diverge northward and southward or say poleward. The surface winds in the name of trade winds blow from subtropical high pressure belts to equatorial low pressure belt in order to replace the ascending air at the equator.

The upper air moving in opposite direction to surface winds (trade winds) is called antitrade. These upper air antitrades descend near 30°-35° latitudes to cause subtropical high pressure belt. These antitrades after descending near 30°-35° latitudes again blow towards the equator where they are again heated and ascend. Thus, one complete meridional cell of air circulation is formed. This is called tropical meridional cell which is located between the equator and 30° latitudes. It may be pointed out that the regularity and continuity of the antitrade wind systems in the upper air has been refuted by a host of meteorologists on the basis of more upper air data being available during and after Second World War.

(2) *Polar front cell* or *mid-latitude cell* : According to old concept surface winds, known as westelies, blow from the subtropical high pressure belt to subpolar low pressure belt (60°-65°). The winds ascend near 60°-65° latitudes because of the rotation of the earth and after reaching the upper troposphere diverge in opposite directions (poleward and equatorward). These winds (which diverge equatorward) again descend near horse latitudes (30°-35° latitudes) to reinforce subtropical high pressure belt. After descending these winds again blow poleward as surface westerlies and thus a complete cell is formed.

According to new concept of air circulation the pattern between 30°-60° latitudes consists of surface westerlies. In fact, winds blow from sub-tropical high pressure belt to subpolar low pressure belt but the winds become almost westerly due to Coriolis force. It may be mentioned that the regularity and continuity of westerlies are frequently disturbed by temperate cyclones, migratory extratropical cyclones and anticyclones. Contrary to the existing view of upper air tropospheric easterly winds in the zones extending between 30°-60° latitudes Rossby observed the existence of upper air westerlies in the middle latitudes due to poleward decrease of air temperature. According to G.T. Trewartha the middle and upper tropospheric westerlies are associated with long waves and jet streams. Warm air ascends along the polar front which is more regular and continuous in the middle troposphere. It may be pointed out that this new concept does not explain the cellular meridional circulation in the middle latitudes.

(3) *Polar cell* involves the atmospheric circulation prevailing between 60° and poles. Cold winds, known as polar easterlies, blow from polar high pressure areas to sub-polar or mid- latitude low pressure belt. The general direction of surface polar winds becomes easterly (east to west) due to coriolis force. These polar cold winds converge with warm westerlies near 60°-65° latitudes and form polar front or mid-latitude front which becomes

the centre for the origin of temperate cyclones. The winds ascend upward due to the rotation of the earth at the subpolar low pressure belt and after reaching middle troposphere they turn poleward and equatorward. The poleward upper air descends at the poles and reinforce the polar high pressure. Thus, a complete polar cell is formed.

Numerous objections have been raised against the concept of tricellular meridional circulation of the atmosphere. The temperature gradient should not be taken as the only basis for the origin and maintenance of cellular meridional circulation because not all the high and low pressure belts are thermally induced. For example, the subtropical high pressure and subpolar low pressure belts are dynamically induced due to subsidence and spreading of air caused by the rotation of the earth respectively. Upper air anti-trades are not uniformly found over all the meridians. If the trade winds are exclusively of thermal origin, then the thermal gradient must be present boldly throughout the tropics but this is not true. At the height of 500 to 1000m in the atmosphere the winds become almost parallel to the isobars which are generally parallel to the latitude. If this is so, the meridional cell of air circulation may not be possible. The pressure and winds in most parts of lower atmosphere are found in cellular form rather than in zonal pattern. These pressure and winds cells are elliptical, circular or semicircular in shape. These evidences (cellular form of air circulation) no doubt contradict the old concept of general pattern of atmospheric circulation but the cellular meridional circulation has not been fully validated.

MONSOON WINDS

1. Meaning and Distribution

The word 'monsoon' is used to indicate the winds in the areas where they change their direction twice each year. In fact, the word 'monsoon' which has been derived from Arabic word 'mausim' or Malayan word 'monsin' meaning thereby 'season' refers to such an atmospheric circulation which reverses its direction completely every 6 months or say during summer and winter seasons. The word 'mausim' was first used by Arab navigators for the winds blowing over the Arabian Sea between Arab and India wherein they blow from north-east to south-west for 6 months during winter season and from south-west to north-east during summer season. On this basis, the word monsoon was applied to all those winds of the globe which had directional change from summer season to winter season and *vice-versa*. In may be pointed out that there are many such places on the globe where there is complete seasonal reversal in the wind direction *e.g.* the region lying between 60°-70° latitudes in the northern hemisphere is characterized by northeast polar winds during winter season and by south-west westerlies during summer season and the Mediterranean regions (30°-40° latitudes) are characterized by westerlies during winter season and north-east trade winds during summer season but these winds are not called *monsoons*. It is apparent that directional change of the winds is not the only criterion of monsoons. In fact, the monsoons are surface convective systems which are originated due to differential heating and cooling of the land and water (oceans) and thermal variations. The regions dominated by monsoon winds are called '*monsoon climatic* regions' which are more developed in Indian subcontinent, south-east Asia, parts of China and Japan. Besides, southern USA, northern Australia, western Africa etc. also represent pseudo-monsoons.

According to Chang-Chia Ch'eng *monsoon is a flow pattern of the general atmospheric circulation over a wide geographical area, in which there is clearly dominant wind in one direction in every part of the region concerned, but in which this prevailing direction of wind is reversed (or almost reversed) from winter to summer and from summer to winter.*

According to Nieuwolt, the word monsoon is used only for wind system where the seasonal reversal is pronounced and exceeds a minimum number of degrees (120 degrees).

2. Concepts of the Origin of Monsoon

The concept of the origin of monsoon is related to thermal and dynamic factors and thus there are two concepts of the origin of monsoon *e.g.* :

(1) Thermal concept, and

(2) Dynamic concept.

(1) Thermal Concept

The thermal concept of the origin of monsoon was first propounded by Halley in 1686. According to this concept the monsoons are the result of heterogeneous character of the globe (unequal distribution of land and water) and differential seasonal heating and cooling of the continental and oceanic areas. During northern winters (winter solstice) when the sun becomes vertical over the tropic of Capricorn in the southern hemisphere high pressure areas are developed over Asia due to very low temperature. Two bold high pressure areas are developed near Baykal Lake and Peshawar. On the other hand, low pressure centre is developed in the southern Indian Ocean due to summer season and related high temperature in the southern hemisphere. Consequently, winds start blowing form the high pressure land areas to the low pressure oceanic areas. These are called *north-east monsoons* or *winter monsoons* (Fig. 3.13). These are dry winds because they come from over the land.

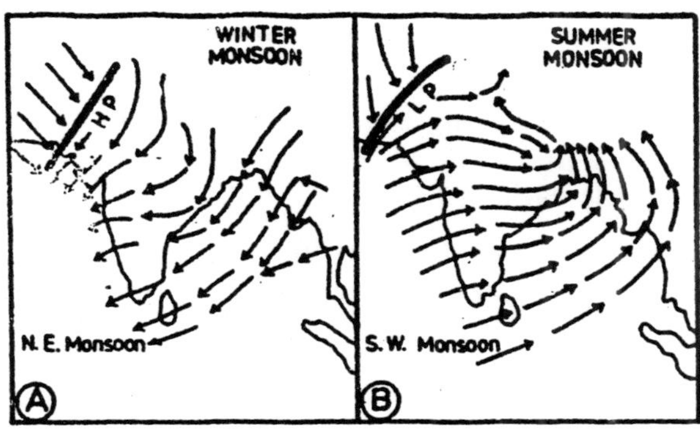

Fig. 3.13 : A-Winter monsoon, B-Summer monsoon.

The aforesaid conditions are reversed at the time of summer solstice when the sun becomes vertical over the tropic of Cancer in the northern hemisphere. Because of high temperature low pressure centres are developed at two places due to the presence of the Himalayas *e.g.* near Baykal Lake and near Multan. Conversely, high pressure centres are developed in the southern Indian Ocean, to the north of Australia and to the south of

Japan because of winter season in the southern hemisphere. Consequently, winds blow from Indian Ocean to Asian continent. These winds while crossing the equator become south-westerly due to *Ferrel's law*. These winds pick up much moisture while passing through the ocean and yield heavy rainfall when obstructed effectively. These are called south-west monsoons or summer monsoons.

(2) Dynamic Concept

A host of scientists have refuted the thermal origin of monsoon and have raised the following objections against the old concept or thermal concept-

(i) If the *lows* developed over the land areas are '*heat lows*' (low pressure centres developed due to high temperature), then they should remain stationary at their places for some time but they are never stationary. There is sudden and widespread shifting in their positions. It, thus, appears that the low pressure centres are not related to thermal conditions, rather they represent cyclonic lows associated with the south-west monsoon.

(ii) The rain producing capacity of monsoon winds is also doubtful. In fact, the monsoon rainfall is associated with tropical disturbances.

(iii) If the monsoons are thermally induced then there should be anti-monsoon circulation in the upper air. It may be pointed out that a few meteorologists have also noticed seasonal variations of winds aloft in the troposphere and stratosphere. Such upper air winds, which change their directions seasonally, are called 'upper air monsoon' or 'aerological monsoon'.

Based on above objections Flohn rejected the thermal concept of the origin of monsoon winds and propounded his new concept which is based on the dynamic origin of monsoons. According to him monsoons are originated due to shifting of pressure and wind belts. Tropical convergence is formed due to convergence of north-east and south-east trade winds near the equator. This is called *intertropical convergence* (ITC). The northern and southern boundaries of ITC are called NITC and SITC respectively (Fig. 3.9). There is a belt of doldrum within the intertropical convergence characterized by equatorial westerlies. At the time of summer solstice (June 21) when the sun becomes vertical over the tropic of Cancer NITC is extended upto 30°N latitude covering south and south-east Asia and thus equatorial westelies are established over these areas. These equatorial westerlies become south-west or summer monsoons. The NITC is associated with numerous atmospheric storms (cyclones) which yield heavy rainfall during wet monsoon months (July to September). Similarly, the north-east or winter monsoon does not originate due to low pressure in the southern hemisphere during winter solstice (southern summer, when the sun becomes vertical over the tropic of Capricorn). In fact, the north-east monsoons are north-east trade winds which are re-established over south and south-east. Asia during northern winter (winter solstice) due to southward shifting of pressure and wind belts and NITC. It is obvious that due to southward movement of the sun at the time of winter solstice the NITC is withdrawn from over south and south-east Asia and north-east trade winds occupy their normal position. These north-east trades, thus, become winter monsoons. Since they come from over the land, and hence they are dry.

3. Origin of Indian Monsoon

The findings of researches conducted in connection with the Indian monsoon after 1950 have revealed that its origin and mechanism are related to the following facts-

(i) The role of the position of the Himalayas and Tibetan plateau as mechanical barrier or as high-level heat source.

(ii) The existence of upper air circum-polar whirls over north and south poles in the troposphere.

(iii) The circulation of upper air jet streams in the troposphere.

(iv) Differential heating and cooling of huge landmass of Asia and Indian Ocean.

Before 1950 the origin and mechanism of Indian monsoon was related to surface air circulation and thermally induced low and high pressures and thus monsoon was considered to be a simple air circulation system but the studies of air circulation in the middle and upper troposphere have shown that the monsoon is a complex air circulation system. High pressure is developed due to extremely low temperature and descent of air from above in the arctic circle over the poles whereas upper air low pressure is developed in the troposphere just above the surface of high pressure area. Thus, upper air circumpolar whirl is developed above the poles wherein winds blow following curved paths in a cyclonic system. In other words, winds blow around upper air low pressure centre in a cyclonic pattern and thus form a whirl. This whirl is called circumpolar whirl. The general direction of air movement over Asia is from west to east. The equatorward winds of this upper air whirl are called *jet streams*.

The jet streams blow in a meandering course. The arctic upper air whirl becomes more prominent and active during winter season in the northern hemisphere (at the time of winter solstice) and thus the upper air westerly jet steams are established in the latitudinal zone of 20°-35°N. Because of equatorward shifting of upper air circum-polar whirl the north intertropical convergence (NITC) is pushed further southward from its natural position. North-east trade winds form the surface air circulation over Indian subcontinent during northern winter (Fig. 35.14A).

The upper air westerly jet streams are positioned in Asia at the height of about 12 kilometres in the troposphere. These jet streams are bifurcated due to the mechanical obstruction of the Himalayas and Tibetan plateau during northern winter. The northern branch blows from west to east in arcuate shape to the north of the Himalayas and Tibetan plateau (Fig. 3.14A) while the southern branch moves from west to east to the south of the Himalayas. It may be pointed out that the main branch of upper-air jet streams follows anti-cyclonic path wherein anticyclonic air circulation is developed to the right of the general flow direction of the jet streams across Afghanistan and Pakistan with the result upper air high pressure is formed over them (Fig. 3.14A, indicated by High) at the height of 10-12km and hence winds descend and settle downward. Conversely, the main branch of jet streams to the south of the Himalayas follows cyclonic arc having anticlockwise air circulation due to mountain barrier. With the result upper air low pressure and cyclonic air circulation are developed over Tibetan plateau (Fig. 3.14A. indicated by Low).

Let us discuss the general conditions during winter and summer seasons. The upper air westerly jet streams are extended upto 20°-35°N latitude due to equatorward shift of upper air north polar whirl during northern winter (October to February). The upper air westerly jet streams are bifurcated into two branches due to mechanical obstructions of the Himalayas and Tibetan plateau. One branch is located to the south of the Himalayas while the second branch is positioned to the north of Tibetan plateau (Fig. 3.14A). Upper air high pressure and anticyclonic (with clockwise air circulation) conditions are developed in the troposphere over Afghanistan and Pakistan. Consequently, the winds tend to descend over north-western part of India resulting into the development of atmospheric stability and dry conditions. Besides, the upper air westerly jet streams also cause periodic changes in general weather conditions because they lie over the temperate low pressure (cyclonic waves) or cyclones which move from west to east under the influence of upper air westerly jet streams across the Mediterranean Sea and reach Pakistan and north-west India. These storms are not frontal cyclones but are waves which move at the height of 200 metres from mean sea level, while at the surface there are north-east trade winds. The arrival of these temperate storms causes precipitation and abrupt decrease in air temperature. The weather becomes clear after they pass away. On an average 4 to 8 cyclonic waves per month reach north-western India between October and April each year. They affect the weather conditions during winter seasons upto Patna.

Now question arises as to why there is no regularity and continuity in the winter cyclonic waves? As stated earlier the upper air high pressure and anticyclonic systems are positioned above the ground surface frequented by cyclonic waves. This

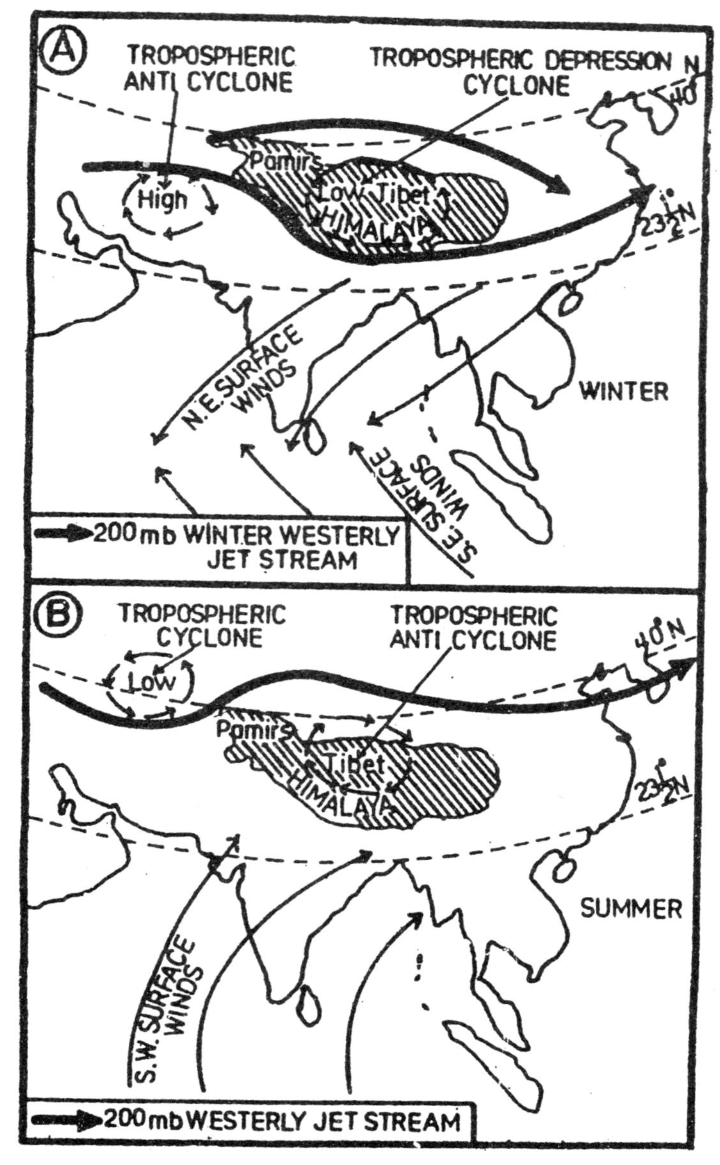

Fig. 35.14: Origin of Indian Monsoon, A-conditions during winter season and B-conditions during summer season.

is why the winds descend and the cyclonic waves located at the height of 200 m from the surface are unable to ascend because the winds descending from the upper air high pressure obstruct them. Simultaneously, the surface trade winds positioned below the cyclonic waves also cool them from below. Consequently, most of the precipitation from these cyclonic waves is orographic in character (the winds rise along the Himalayas and yield precipitation due to cooling and increase in relative humidity). On an average, most parts of India remain dry during winter season except Tamil Nadu coast which receives much rain-fall during October-November.

After vernal equinox (21 March) sun moves northward and becomes vertical over the tropic of Cancer (at the time of summer solstice, 21 June) with the result the polar surface high pressure is weakened and upper air circum-polar whirl which extended upto 200-350N latitudes during winter season shifts northward due to which upper air westerly jet streams are also withdrawn and shift northward. Thus, the dynamic force of the polar whirl is weakened. Consequently, the upper air circum-polar whirl becomes unable to maintain the southern branch of the westerly jet streams (to the south of the Himalayas, Fig. 3.14A) and thus they (jet streams) shift to the north of the Himalayas and Tibetan plateau (Fig. 3.14B). Final withdrawal of upper air streams from over India is completed by middle of June.

Low pressure areas are developed at the ground surface in north-west Pakistan and north-west India due to intense heating of ground surface during April-May. But so long as the position of upper air jet streams is maintained above the surface low pressure (to the south of the Himalayas), the dynamic cyclonic conditions persist over Afghanistan, north-west Pakistan and north-west India. The winds descending from the upper air high pressure obstruct the ascent of winds from the surface low pressure areas, with the result the weather remains warm and dry. This is why the months of April and May are dry inspite of high temperature and evaporation. It may be pointed out that monsoon arrives in May in Myanmar but north-west India remains dry. Upper air low pressure is formed to the east of the eastern limit of the Himalayas due to upper air jet streams, with the result the winds coming from south in Myanmar are forced to ascend and yield copious rainfall. The Myanmar monsoon also affects Bangladesh and adjoining Indian territory which receives premonsoon rainfall.

The upper air westerly jet stream (southern branch of winter jet stream) is withdrawn from over India by the middle of June (Fig. 3.14B), reason has already been explained above. Now, the jet stream is positioned to the north of Tibet and the trajectory of its flow becomes opposite (Fig. 3. 14B) to the flow curvature during winter season (Fig. 3.14A). The flow path of upper air westerly jet stream becomes cyclonic curve (anti-clockwise movement of free air) over Iran and Afganistan due to which dynamically induced low pressure is formed in the upper air (in the troposphere, called as tropospheric low or cyclone, Fig. 3.14B) and thus cyclonic conditions dominate the upper atmosphere. It may be remembered that there is high pressure and anticyclonic conditions during winter season in the areas of summer upper air tropospheric low pressure and cyclonic condition. This upper air low pressure is also extended over Pakistan and north-west India. There is already thermally induced low pressure at the ground surface located below the upper air low pressure. Consequently, the surface warm winds rise upward. The ascent of surface warm air is further accelerated because the upper air low pressure sucks the air from the ground surface. This mechanism causes sudden burst of south-west monsoon.

It may be remembered that during northern summer there is winter season in the southern hemisphere, with the result southern polar whirl is more developed and is extended upto the equator. Consequently, the intertropical convergence (ITC) is pushed to the north of equator. Because of the push factor of the southern polar whirl the south-east trade winds are forced equatorward and while crossing over the equator they become south-westerly due to coriolis force (deflective force caused due to the rotation of the earth) and rush towards India. It may be pointed out that rapid advance of inter-tropical convergence northward is because of the push factor of the southern circumpolar whirl and not because of sucking by the thermally induced surface low pressure over north-west India.

No doubt, this surface low pressure accelerates the advance of intertropical convergence northward. Intertropical convergence is characterized by dynamically induced waves and not by frontal cyclones. These dynamically induced waves after coming over India become cyclone vortices. The summer monsoon rains of India result from these cyclonic vortices. In other words, the development of cyclonic vortices is followed by wet weather while their occlusion causes dry weather which continues till new cyclonic vortex is formed.

There is much spatial and temporal variation of monsoon rainfall in India. Topographic factor plays a major role in such variation. For example, the Arabian Sea Branch of south-west monsoon rises abruptly after being obstructed by the Western Ghats and yield heavy rainfall while the regions located to the east of the Western Ghats receive meager amount of rainfall because the winds descend along the eastern slopes and thus are wanned due to which relative humidity decreases and aridity increases. Such regions of low rainfall are called 'rain-shadow regions'. The Himalayas affect the Bay of Bengal Branch of south-west monsoon in two ways *e.g.* (i) the air ascends due to obstruction of the mountain and yields heavy rainfall, and (ii) the obstruction by the Himalayas causes channel effects due to which winds blow westward along the mountains. Consequently, the monsoon reaches northwestern India through the Ganga valley. Inspite of strong surface low pressure over Rajasthan and adjoining Pakistani territory the rainfall is minimum. Generally, the lowest amount of rainfall over north-west India is related to the parallel position of the Aravallis to the Arabian Sea Branch of south-west monsoon but the real cause is related to the depth of monsoon drift which depends on the position of upper air dynamic anticyclonic condition above surface low pressure.

This upper air high pressure obstructs the upward movement of surface winds. Whenever this upper air high pressure shifts westward, monsoon winds rise rapidly and yield heavy rainfall even in Rajasthan.

MONEX

The Monsoon Expedition (MONEX), under the joint auspices of Russian and Indian scientists, organized in 1973, revealed certain new information about the weather conditions over the Indian sub- continent. The meteorologists, based on the data received from MONEX, propounded that the Indian monsoons are related to the heating and cooling of Tibet Highland (having the widths of 1000 km and 600 km in the east and west respectively, the length of 2000 km and the height of 4000m). The mechanism of the effects of the Tibetan plateau and the Himalayas has already been explained above.

LOCAL WINDS

1. Land And Sea Breawa

Land and sea breezes, representing a complete cycle of diurnal winds, are, in fact, monsoon winds at local scale because they change their direction twice in every 24-hour period. These local diurnal monsoon winds very commonly known as land and see breezes are found in the coastal areas wherein sea breeze blows from sea to land, during day time and land breeze moves from land to sea during night due to differential heating and cooling of land and water.

Sea Breeze

Land is heated more quickly than the adjacent sea during day time, with the result the warm air over the adjacent land expands and thus low pressure is developed while high pressure is developed over adjacent sea. The pressure gradient causes circulation of relatively cool air from sea to adjacent land (Fig. 3.15). Sea breezes begin to flow usually between 10-11 a.m. and become most active in early afternoon usually between 1-2 p.m. with maximum velocity ranging between 10 to 20 kilometres per hour and are terminated by 8 p.m. at night. The average depth of sea breeze system ranges between 1000-2000 metres in the coastal regions of the tropical areas while its depth is between 200 and 500 m near the lakes. The cooling effect of sea breezes reaches 50 to 65 km inland in the tropical regions while 15 o 50 km in the middle latitudes. The velocity of these winds varies spatially *e.g.* the velocity varies from 25 to 50 km per hour in the temperate areas while some times sea breezes become stormy in the tropical areas. Sea breezes have cooling effects on the coastal land as the temperature drops by 5°C to

10°C, with the result weather becomes pleasant. Sea breezes are most active during summer season.

Land Breeze

After sunset the sea breezes are weakened because the day time low pressure over land is weakened due to rapid loss of heat through outgoing longwave radiation from the land. Consequently, the position of day time high and low pressure is reversed. Now high pressure is developed on land against low pressure on the adjacent sea with the result air starts moving from land to sea during night (Fig. 3.15B). Land breezes are comparatively weaker than sea breases. These are dry winds.

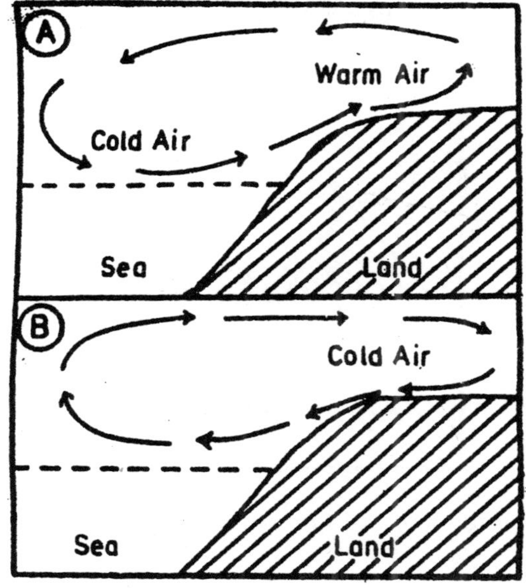

Fig. 3.15 : A = Sea breeze, B = land breeze.

2. Mountain and Valley Breezes

Mountain and valley breezes also known as upvalley (during daytime) and downvalley (during night) breezes are, in fact, local as well as diurnal (periodic winds the directions of which are reversed during 24 hours). The slopes and valley floors in the mountainous regions are more heated through insolation during daytime than the free atmosphere at the same elevation. Consequently, the warm air moves upslope (upward). This upslope moving breeze during daytime is called valley breeze (Fig. 35.16). Valley breezes reach mountain peaks and yield precipitation through cumulus clouds. The valley slopes and upper parts are cooled due to loss of heat through outgoing longwave radiation and thus cool air descends through the valley slopes. Such wind is called *down valley* or *mountain breeze*. The mountain breezes cause inversion of temperature in the valleys. This is why the valley floors are characterized by frost during night while the upper parts are free from frost in cold areas.

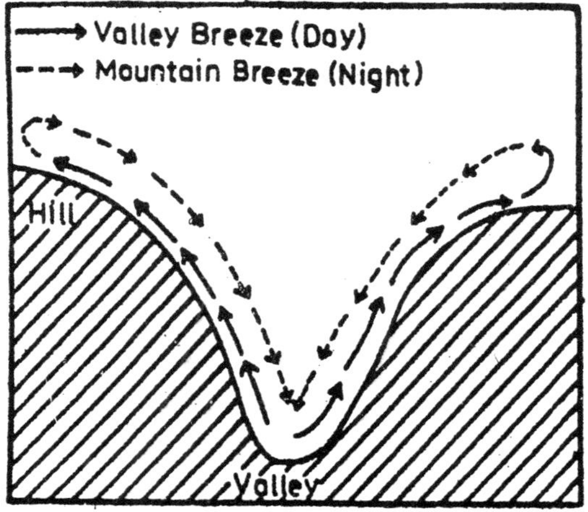

Fig. 3.16 : Mountain and valley breezes.

3. Chinook And Foehn

Warm and dry local winds blowing on the leeward sides of the mountains are called *chinook* in the USA and foehn in Switzerland. These local vertical winds are of cyclonic origin and largely influence the weather conditions of the affected areas locally. The winds associated with the cyclones after descending through the eastern slopes of the Rockies become warm and dry and thus give birth to chinook. The winds ascend through the western slopes of the Rockies mountains and thus are cooled at the dry adiabatic rate of 5.5°F per 1000 feet (10°C per 1000 metres). These winds after reaching higher height become saturated (due to lowering of temperature and hence, increase of relative humidity) and yield precipitation. The latent heat of condensation released after precipitation is added to the ascending winds, with the result the temperature of the ascending winds decreases at the moist adiabatic rate of 3°P per 1000 feet (Fig. 3.17). The westerly winds after crossing over the Rockies descend through the eastern slopes and thus are heated at the dry adiabatic rate of 5.5°F per thousand feet. These warm and dry winds after reaching the foothill zones of the eastern slopes of the Rockies are called *Chinooks*.

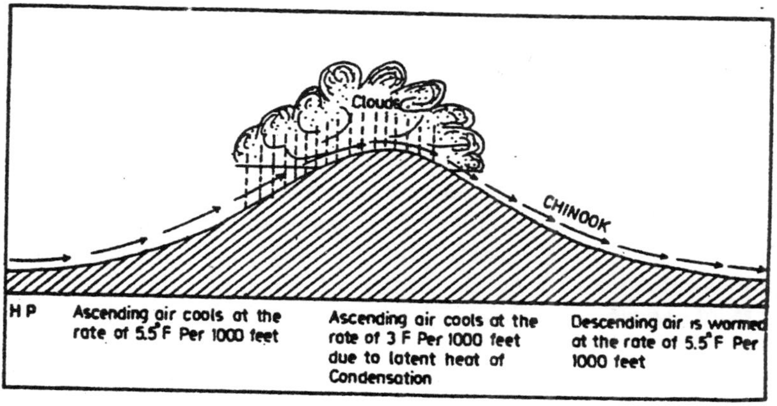

Fig. 3.17 : Origin of chinook winds.

Chinook winds are more common during winter and early spring along the eastern slopes (leeward side) of the Rocky Mountains from Colorado (USA) in the south to British Columbia (Canada) in the north. Normally, the actual temperature of chinooks is 40°F (4.4°C) but the actual temperature of the affected areas is below freezing point during winter months. Thus, chinook becomes very warm in comparison to prevailing subzero temperature. The rise of temperature in the affected areas by 40°F (4.4°C) after the arrival of chinooks within 24 hours is not unusual.

Sometimes, temperature rises by 30° to 40°F within few minutes. Thus, snow present on the ground surface melts away due to sudden rise in temperature as if by magic. This is why chinook is also called as 'snow eater'. This is the impact of chinooks that green pastures are open in a narrow strip along the eastern slopes of the Rockies even during winter season. The rise in temperature due to chinooks also helps in early sowing of spring wheat in the USA.

A warm and dry wind similar to chinook is called '*foehn*' along the northern slopes of the Alps mountains. These are more common during spring and autumn in Switzerland. The weather becomes quite pleasant in the valleys due to melting of snow because of increase in temperature by 40°F after the arrival of foehn winds. This is why valleys of Switzerland are called '*climatic oasis*' during winter season. These winds help in early sowing of spring wheat, ripening of grapes and check autumn frost.

4. Harmattan

The warm and dry winds blowing from north-east and east to west in the eastern parts of Sahara desert are called harmattan. These winds become extremely dry because of their journey over Sahara desert. While blowing over Sahara desert these winds pick up more sands mainly red sands. The western coast of Africa is warm and moist and hence the weather becomes unpleasant because the weather conditions characterized by high temperature and high relative humidity become injurious for human health. The weather becomes suddenly dry and pleasant at the arrival of harmattan as the relative humidity of the air is remarkably reduced due to high temperature and hyperaridity of harmattan. This is why harmattan is known as 'doctor' in the Guinea coastal area of western Africa. In fact, harmattan is very dusty and stormy wind blowing with so gusty speed that trees are uprooted. These winds are usually associated with dust storms resulting into marked reduction in the visibility. Harmattan becomes more vigorous during summer months. In fact, harmattan is a special type of north-east trade wind. It becomes extremely warm wind because of hot and dry desert of Sahara. Similar warm, dry, very strong and dustladen winds are called '*brickfielder*' in Victoria province of Australia, '*blackroller*' in the Great Plains of the USA, '*shamal*' in Mesopotamia and Persian Gulf, and '*norwester*' in New Zealand.

5. Sirocco

Sirocco is a warm, dry and dusty (full of sands) local wind which blows in northerly direction from sahara Desert and after crossing over the Mediterranean Sea reaches Italy, Spain etc. Sirocco becomes very strong and active at the time of the origin of cyclonic storms over the Mediterranean Sea. It becomes extremely warm and dry while descending through the northern slopes of the Atlas mountain. There are different local names for sirocco in Africa *e.g. khamsin* in Egypt (UAR), *gibli* in Lybia, *chilli* in Tunisia etc. The warm and dry dusty winds in the Arabian Desert are called '*simoom*'. Sirocco, while passing over the Mediterranean Sea picks up moisture and yields rainfall in the southern part of Italy where the rain associated with sirocco is called '*blood rain*' because of fallout of red sands with falling rains. It may be remembered that sirocco while blowing through Sahara Desert picks up red sands which settle down with rains in south Italy. It is apparent that sirocco is very much injurious to agricultural and fruit crops.

6. Mistral

Mistral is a cold local wind which blows in Spain and France from north-west to south-east direction. These winds are more common and effective during winter season because of development of high pressure over Europe and low pressure over Mediterranean Sea. They become extremely cold when they blow through central plateau and descend into Rhone valley on the southern coast of France. While blowing through the narrow valley of Rhone they become stormy northerly cold winds. The aver age velocity of mistral is 56-64 km per hour but sometimes it becomes 128 km per hour. These stormy winds adversely affect airflights. The arrival of mistral causes sudden drop in air temperature to below freezing point.

7. Bora

Bora is an extremely cold and dry north-easterly wind which blows along the shore of the Adriatic Sea. Bora becomes more effective in north Italy where it descends through the southern slopes of the Alps and blow in southerly direction. The velocity of bora ranges between 128 km and 196 km an hour. Unlike mistral, bora is relatively moist wind because it picks up moisture while coming from over the Adriatic Sea.

9. Blizzard

Blizzard is a violent stormy cold and powdery polar wind laden with dry snow and is prevalent in north and south polar regions, Siberia, Canada and the USA. The visibility becomes remarkably low because of snow and ice crystals. The velocity ranges between 80-96 km an hour. The arrival of these winds causes sudden drop in air temperature to subfreezing level, thick cover of snow on the ground surface and onset of cold waves. These winds reach the southern states of the USA because of the absence of any east-west mountain barrier. They are called '*norther*' in the southern USA and 'burran' in Siberia.

OTHER LOCAL WINDS

Purga—a snow-laden cold wind in Russian Tundra.

Bise—an extremely cold wind in France.

Levanter—a strong easterly cold wind in southern Spain.

Pampero—a northwesterly cold wind in the 'pampas' of S. America.

Norwester—a warm, dry and gusty wind in New Zealand.

Santa Ana—a warm and dry wind in the USA.

Yamo—a warm and dry wind in Japan.

Zonda—a warm wind in Argentina.

Tramontane—a warm wind in central Europe.

MEASUREMENT AND RECORDING

Atmosphere circulation has a definite weight and so exerts pressure upon the earth. The weight of air in a vertical column extending from the upper limit of the atmosphere to the earth's surface exerts a pressure which averages about 1 kg per sq cm (14.7 lb per sq in). This is, of course, more or less in balance with the force of gravity.

Instruments

The weight of this vertical column of air is measured by a barometer, first done by Torricelli in 1643; in effect he balanced the weight of a column of mercury in a glass tube, sealed at one end, against the pressure of a column of the atmosphere of equal cross-section. Mercury, the heaviest known liquid, is used because a tube of water, for example, would need to be 10 m (34 ft) in length. Modern mercury barometers, such as the Kew or Fortin pattern, have various refinements and adjustments for applying corrections, and vernier scales for making exact readings. These barometers are read in terms of the length of the column of mercury, either in inches or millimetres; 29-92 in (the average sea-level pressure over the world) is equivalent to 760 mm.

During this century an alternative and more logical system of pressure measurement has been evolved, although the mercury column is still commonly used. Physicists use a unit called the *dyne*, the force which, when applied to a mass of 1 gramme for 1 second, will produce a change in velocity of 1 cm per second. A *bar*, representing a pressure of a million dynes per sq cm, is now the standard unit of pressure measurement, divided into a thousand *millibars (mb).* There is no universal formula for conversion, except at a constant temperature and latitude; at 45°N latitude, at sea-level, and at a temperature of o°C, 1 bar equals 29.5306 in or 750.1 mm of mercury, or 30 in of mercury is equivalent to 1015.9 mb. The average sea-level pressure of 29.92 in is equivalent to 1013.25 mb.

In addition to the mercury barometer, other pressure-recording instruments have been devised. The aneroid barometer comprises essentially a metallic box, almost exhausted of air, the sides of which are flexible so that they can expand and contract with the external air pressure. These movements are connected by a spring to a needle which moves round a calibrated circular dial. A barograph operates on a similar principle, with the spring connected to an inked pen which makes a continuous record on a chart fixed to a revolving drum. These aneroid recorders are neither as accurate nor as reliable as the mercury patterns.

Corrections Applied to Barometer Readings

Corrections are applied to recorded pressures in order to attain comparable standards. These corrections are made for

(i) Latitude,

(ii) Temperature, and

(iii) Altitude.

(i) Latitude : The earth is not a perfect sphere, and therefore the force of gravity varies according to the latitude, attaining a maximum at the Poles and a minimum at the Equator. Readings for precise comparisons are therefore standardized to 45° latitude.

(ii) Temperature : Errors in pressure readings are introduced as a result of the different coefficients of expansion of both the mercury column and the metal scale. Tables are supplied with each instrument to enable the correction to be made; the bar is officially standardized at 0°C (32°F).

(iii) Altitude : There is a considerable change of pressure with altitude, since the proportion of the overlying atmosphere decreases with height. A very rough indication of the pressure decrease is about 34 mb (i in of mercury) for every 270 m (900 ft), but this applies only to the lowest 1000 m, above which the rate becomes less. From sea-level to 600 m (2000 ft) the decrease is about 4 per cent for each 330 m (1000 ft); from 600 m to 1500 m (5000 ft) it is about 3 per cent; and from 1500 m to 3000 m (10,000 ft) it is 2-5 per cent. At a height of 24 km (15 miles) the rate of decrease becomes virtually imperceptible. If the pressure is 1016 mb (30 in) at sea-level, it will be about 540 mb (16 in) at the top of Mont Blanc; an aneroid calibrated in feet and/or metres is sometimes carried by mountaineers. The same principle is used in the altimeter of an aircraft.

The physiological results of reduced pressure with altitude are shown by the difficulties of acclimatization in high mountaineering. Indeed, it is unlikely that Everest will be climbed without the help of oxygen from cylinders. Sir Edmund Hillary and Tensing Norkey, who reached the summit of Everest on 29 May, 1953, each used an open-circuit oxygen apparatus weighing over 14 kilograms, from which they inhaled 3 litres of oxygen per minute during the final stages of the ascent.

Pressure Maps

Pressure maps are constructed on the same principles as temperature maps. Stations are located, the mean pressure values for them are plotted, and lines of equal barometric pressure, known as isobars, are interpolated. These values are reduced to sea-level; as the fall in pressure with altitude is so rapid, isobars (even more than isotherms) would simply follow the contours. Sea-level variation of pressure is relatively small, ranging (as far as mean values are concerned) only from about 980 mb (29 in) to 1070 mb (31.5 in), whereas it can fall to 540 mb (16 in) in the French Alps, and still lower in the higher ranges. The lowest barometric reading for a place near sea-level in Britain was 925-5 mb (27.33 in) in Perthshire on 26 January 1884; this was associated with a severe storm which caused widespread damage. The world's highest recorded sea-level barometric reading was 1079 mb in Russia in 1900, the lowest 877 mb in the Pacific Ocean in 1958.

There are various kinds of pressure map. One is the synoptic chart or daily weather map, which depicts the isobars at a moment in time. Most countries have a central office to which observers send periodic reports, in an internationally accepted code, by teleprinter. The Central Forecasting Office in Great Britain is at Bracknell, where the weather map is prepared, and from which information is sent to airfields, and to international meteorological offices.

These synoptic charts are not merely drawn for surface conditions. In recent years high-altitude meteorological research has made it possible to construct upper-air synoptic charts, using radio-sonde and rawinsonde ascents. These, known as contour-charts, show the height of standard pressure surfaces. They are of great value in high-altitude aviation, as well as helping the forecaster to predict surface weather conditions.

Another type is the *average pressure map* on a continental or world scale, which portrays mean pressures calculated over a long period of time. While such a map represents only a broad generalization of conditions, it is helpful because it brings out the patterns and the salient features of pressure distribution over continent and ocean, season by season.

A useful type of chart shows the alteration of pressure over a period of time by means of isallobars, lines of equal pressure change. When these are plotted, their patterns may reveal how a pressure system is developing and moving, thus assisting weather forecasting.

WIND

Wind Force and Velocity

The measurement and recording of wind involves both the direction from which it is blowing and its velocity. Until 1949 a variety of codes was in use; in British returns direction was expressed in terms of the cardinal points of the compass, force in terms of the *Beaufort Scale* used for a century and a half and based on personal observation, both on shore and on ship. Thirteen 'forces' are used, from 0 (Calm) to 12 (Hurricane), related to descriptions of their effects and their estimated velocity at 10 m above the ground. Thus Force 4 (moderate breeze) of 13 to 18 knots was distinguished at sea by 'smacks carry all canvas', and on land by 'raises dust and paper; small branches move'. Force 8 (fresh gale) of 39–46 knots was similarly described as 'smacks make for harbour' and 'breaks twigs and branches off trees'. Velocities of over 160 km (100 miles) ph are occasionally reached on the west coast of Britain. During the gales of 16 February 1962 a gust of 285 km (177.2 miles) per hour was recorded at an R.A.F. station on the Shetland island of Unit, a British record. Since 1949 most countries have adopted the recommendations of the International Meteorological Organization, using a standard code of velocity in km per hour and giving wind directions as true bearings. Velocity and direction are given in average terms, though usually there is a sequence of lulls and gusts (*i.e., turbulence*) of great importance in aviation.

Instruments

Apart from direct observation and estimation, several wind-recording instruments can be employed. A commonly used instrument is the *cup-anemometer*, consisting of an upright, at the top of which are three horizontal spokes at an angle to each other, with cups fixed at their ends. These arms rotate horizontally as the wind blows, so moving a central rod which transmits the movement to a calibrated dial; some types generate a recorded voltage.

The *Dines pressure tube* has the merit of being a self-recording instrument which provides a continuous record of both direction and velocity. It consists of a vane placed in a prominent position, which has a shaft going down to the recording unit. The front end of the vane has an opening facing into the wind, and communicates the resulting increase of pressure to the recording unit. This provides a precise indication of velocity, while the movement of the vane records changes of direction. Both are recorded as ink lines traced on a revolving drum. This instrument is now obsolescent.

Records of wind in the upper air were formerly made by releasing small balloons with a fixed rate of climb, and observing their trends by means of a theodolite. The horizontal drift was measured, and some estimate made of the wind force at different levels. This method is slow and inaccurate, and has been superseded at modern meteorological stations by a balloon with a radar target, which can be followed accurately by the observer even when it is cloudy.

Presentation of Wind Records

Observations of wind direction and velocity are usually calculated on a percentage basis. Detailed tables give this information for winds from different directions, for the various velocities and also for calms.

Wind Maps

The direction to which they are blowing. Force is indicated either by a number of 'feathers' on the shaft of the arrow to correspond to the code in use, or by arrows of different thicknesses. The mean movement of air, representative of different seasons, may be shown by 'stream-lines', which take the form of elongated arrows curving to denote changes in direction.

A striking depiction of wind is by means of a *wind-rose,* of which there are many varieties. Basically it consists of radiating rays, drawn proportional in length to the mean percentage frequency of winds from each cardinal direction; calms are stated at the centre of the rose. Sometimes part of each ray may be thickened to indicate the frequency of winds above a certain critical force. In other types, twelve radiating rays represent months for the cardinal directions.

WIND SYSTEMS

The relation of wind and pressure Wind may be defined as the movement of air from high to low atmospheric pressure in a virtually horizontal plane. Its direction and strength is the result of four factors:

(i) the barometric gradient,

(ii) the Coriolis force,

(iii) the centrifugal force and

(iv) friction.

(i) *The barometric or pressure gradient :* Regions of high pressure form centres from which winds tend to blow outward, that is, they are areas of *divergence*. Conversely, regions of low pressure are the foci of winds, that is, areas of *convergence*. When the difference in pressure between two neighbouring points on the earth's surface is great, there is said to be a steep pressure gradient, and on a pressure map the isobars are close together; isobars and pressure maps are thus analogous to contours and relief maps. In such a case the resulting wind is strong. When, however, the barometric gradient is gentle, winds are light.

The result is that air tends to move from areas of high pressure to those of low. The succeeding factors, however, considerably modify this generalization.

(ii) *Coriolis force :* The effect of the Coriolis force—the deflecting component produced by the rotation of the earth—may be most clearly stated in terms of *Ferret's Law*. A body moving on the surface of the earth is deflected to its own right in the northern hemisphere, and to its left in the southern hemisphere, as a result of the earth's rotation. This has an effect on ocean currents, on tidal movements on driftwood which tends to collect along a right-hand river bank in the northern hemisphere, and on rifle bullets, which are slightly deflected. The force is proportional to the speed of the moving object, and it varies with the latitude, being zero at the Equator and at a maximum at the Poles.

When a balanced condition develops between the forces exerted by the pressure gradient in one direction and the Coriolis force in the opposite direction, a steady wind will blow. This is known as a state of *geostrophic balance*, and the air movement as *geostrophic flow* or as a geostrophic wind, in a direction parallel to the straight isobars.

(iii) *Centrifugal force :* When an air-stream moves on a curved course, as in a pressure system with closed or curved isobars, it is subject to centrifugal force acting outwards from the centre of curvature. A *gradient wind* develops as a result of the balance between the pressure gradient on the one hand, and the Coriolis and centrifugal forces on the other. The air thus tends to travel along the isobars, in the northern hemisphere clockwise around the highs and anti-clockwise around the lows, while in the southern hemisphere it would tend to move anti-clockwise around the highs and clockwise around the lows. This relation of wind to atmospheric pressure was put forward by a Dutch scientist. Buys Ballot, in the middle of the nineteenth century; if you stand with your back to the wind in the northern hemisphere, pressure is lower on your left than on your right, and the reverse applies in the southern hemisphere.

(iv) *The effect of friction :* Winds approaching pure geostrophic flow are found in the upper atmosphere. Near the ground, however, friction tends to reduce the power of the deflection due to rotation, so the net result is that the air crosses the isobars at a slight angle from high pressure to low pressure. This effect is greater over the land than over the oceans, and less with increasing altitude.

With these concepts in mind, the major wind systems of the earth can be described. Just as the theoretical distribution of pressure on a homogeneous globe was a useful initial step in understanding, so it is of considerable help to describe the wind systems which would be associated with this idealized arrangement of pressure belts and then examine the actual modifications.

(i) *The Trades :* The Trade Winds blow from the Subtropical highs towards the Equatorial low, but as they are deflected by the rotating earth they blow from a north-easterly direction in the northern hemisphere (the North-east Trades) and from a southeasterly direction in the southern hemisphere (the South-east Trades). The Trades system moves with the other pressure and wind belts north and south with the seasons through about 7° of latitude. In the northern hemisphere winter the North-east Trades cover roughly the zone between Cancer and a few degrees north of the Equator, whereas in summer they occupy a belt farther north, roughly from about 30°N to about 10°N, and the system of the South-east Trades moves accordingly northward.

The Trades are noted for their constancy of force and direction, particularly over the eastern sides of the oceans, though they are markedly stronger in winter than in summer. But many interferences by local pressure disturbances occur on the western sides, and also near the Equator; here the complicating effects of their convergence, together with increased vertical convection lifting, are experienced P. R. Crowe computed the percentage constancy of these winds for each of the 990 sea areas (of 5° of latitude and longitude) between 45° N and S; over much of the oceans they have a constancy of 70 per cent, in the east-centres of the oceans of over 90 per cent. Their force is for the most part from 16 to 30 km (10 to 20 miles) per hour, in places 24 to 40 km (15-25 miles) per hour, while calms are extremely rare, and clear sunny conditions are usual.

The name Trade Wind comes from the phrase 'to blow trade', *i.e.*, to blow steadily in a constant direction. The name originally had nothing to do with trading, although in sailing-ship days these winds certainly furthered trade.

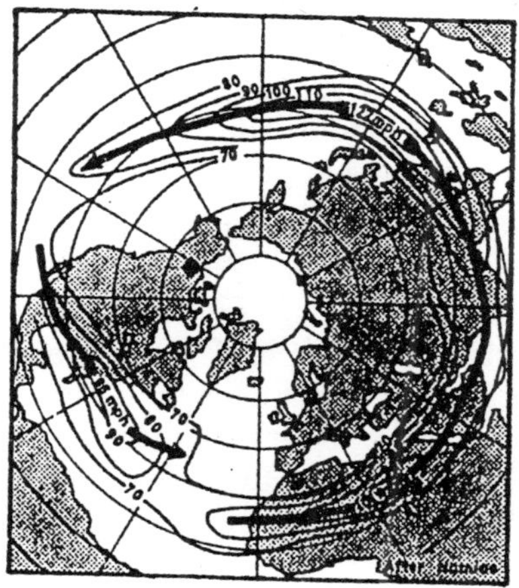

Fig. 3.18 : The air circulation in the troposphere

Between the two Trade Wind systems is found what is known as the Intertropical Convergence Zone; part of this is occupied by an area of calms or light winds, the Doldrums. However, it has recently been discovered that in the 'summer' hemisphere, and especially over the continents (where thermal heating is pronounced), a zone of westerly winds intervenes between the Trades; these are called the *'Equatorial Westerlies'*.

(ii) *The Mid-latitude Westerlies* : The Westerlies comprise the air-flow from the Subtropical highs to the Subpolar lows, from about 35° to 60°; these belts move north and south with the seasonal change of the pressure belts. They consist essentially of two separate circum-polar vortices, increasing in both speed and constancy with altitude, and complicated by jet streams (below) in the upper troposphere. Within the belts both wind direction and velocity are far from constant, the result of local pressure systems, but a westerly component is predominant; the most frequent winds in the northern hemisphere are from the south-west, hence the name 'South-westerly Variables' or 'disturbed Westerlies'.

In the southern hemisphere, where there is continuous open ocean from 40° to 60° S, the Westerlies blow with strength and constancy and are characterized by gales, stormy seas, overcast sloes and damp, raw weather. Seamen call them 'the Brave West Winds', and the latitudes in which they blow 'the Roaring Forties'.

(iii) *The Polar Easterlies* Cold air tends to move equatorward from the Polar highs. This movement is pronounced in the southern hemisphere, where both the high over the Antarctic plateau and the Subpolar low over the Southern Ocean are clearly defined. As the air moves equatorward it is deflected to the left, and so a belt of winds results, spiralling from the polar region in an easterly direction, forming a polar vortex.

In the northern hemisphere, pressure and wind conditions are so complicated that these polar winds are extremely irregular. Sometimes a large part of North America or Asia is affected by the polar winds which spread southward.

Upper-air movement The winds described so far are mainly surface air-currents. In the upper troposphere there appears to be a general movement of air eastward (hence the name 'Upper Westerlies'), largely a result of the upper air-pressure distribution and of the thermal gradient from Equator to Poles. The air-streams move in a series of waves or loops at a height of about 10 km (6 miles), a kind of 'circum-polar whirl'. In these loops originate the depressions of mid-latitudes. The maximum velocities of these westerlies appear to be found at about 30° latitude, the so-called Subtropical Jet Stream discussed below. These velocities are greater than at sea-level because of the reduced density of the atmosphere at these altitudes. Between the high-pressure cells at about 15° North and South lies a zone of upper winds, though of low velocity, known as the 'Equatorial Easterlies'.

The Jet Streams : Towards the end of the War of 1939-45, American bombers in the Pacific area encountered powerful currents of air at high altitudes flowing from west to east; an aircraft flying westward could at times hardly hold its own. The term 'jet stream' was coined, rather a misnomer since it is neither a jet nor a stream, but it has been accepted. The Polar Front Jet occurs in middle latitudes at 9000-12,000 m (30,000–40,000 ft), more or less at the tropopause; it fluctuates, however, both north and south and as low as 3000 m (10,000 ft). The more continuous and persistent

Subtropical Jet occurs at about 12,000 m (40,000 ft), in effect vortices associated with the mid-latitude cell. Others include the *Arctic Jet* which has been traced across Alaska and Canada at 7600 m (25,000 ft), and the recently discovered *Polar night Jet* above the Arctic Circle in the lower stratosphere. An increasing number of observations indicate that in summer the mean speed of the jets is about 90.110 km per hour (50–60 knots), but in winter it is considerably higher (170–220 km per hour, 90–120 knots), and on occasions an exceptionally strong jet has been experienced, once as much as 460 km per hour (250 knots). These are of such importance to aviation, especially over the expanses of the Pacific, that every day a weather reconnaissance plane flies a 3700 km (2300 miles) route from Honolulu, another from Fairbanks, Alaska, and a third from San Francisco; their reports enable the day's jets to be plotted.

In addition to these global jet streams, there are others of a much more local and transient character, a kind of 'fast river of air' of almost tubular cross-section about 160 km (100 miles) in width and 3–8 km (2–5 miles) deep, extremely turbulent at its margins though smoothly flowing within; east-bound aircraft where possible 'ride the Big Wind' across the Atlantic and Pacific to save time and fuel.

The jet streams have considerable significance in connection with surface weather conditions, though this is by no means understood. In general, the more wavy and 'scalloped' the path of the jet, the more numerous and active are local pressure systems near the surface of the earth. In the U.S.A. a distinct relationship has been discovered between jet activity and the incidence of tornadoes when the Polar Front Jet bends south over the Mississippi valley storms may be expected, and the Severe Local Storm Forecast Center at Kansas City, run by the U.S. Weather Bureau, may issue a warning. A dip of the jet toward the ground may mean torrential rain or hailstorms. The position and nature of the jets appear to have important meteorological results, especially on *frontogenesis* and on the 'bursting' of the Monsoon.

Fig. 3.19 : The Subtropical Jet Stream.

Interruptions to the Planetary Wind Systems

The planetary system of winds holds good to a large extent for most of the southern hemisphere, where the interrupting effects of the land-masses are not serious. But in the northern hemisphere and in that part of the southern hemisphere which lies north of 30°S, the planetary wind systems are considerably modified, as are those of pressure already described. These modifications include:

(i) The seasonal reversal of pressure and winds over the land-masses and neighbouring oceans, usually called the monsoons,

(ii) The formation of 'air circulation wheels', moving around the 'cells' of both high and low pressure formed by the interruption of the planetary pressure belts, and

(iii) The movement of air around more local small-scale pressure systems.

The Monsoons

The word 'monsoon' is derived from an Arabic word mausim, meaning season. It has already been seen that the great differences in seasonal temperature over the continents produce striking reversals in pressure conditions, particularly in central Asia. In summer a low develops over north-western India, which is so intense that it supersedes the Equatorial low. As a result an uninterrupted pressure gradient develops from the Subtropical high in the southern hemisphere across the Equator to north-western India. The South-east Trades cross the Equator, and, as a result of the influence of the rotation of the earth, become the South-west Monsoon over peninsular India, and Burma, Indonesia, China and Japan experience winds from a south-easterly direction, blowing towards the heated continent.

In winter the great cold of central Asia results in the formation of an intense high, from which cold, dry winds blow outward from the north-west over Japan and China, and from the north or north-east over south-eastern Asia. This outward flow is strengthened by great air-masses subsiding from higher altitudes. The implications of this seasonal reversal are enormous six monsoonal climatic types are distinguished.

This simple and traditional explanation of the Asiatic monsoons, merely in terms of the marked seasonal differential heating and cooling of land and sea is, however, far from adequate. In addition, the poleward shift of the hemispheric wind-belts in summer, both near the surface and in the upper troposphere, is involved. The Asiatic monsoon is considerably affected by the interrupting effect of the lofty and extensive Plateau of Tibet on the Upper Westerlies. In summer these shift north of the Plateau, emphasized by the north-ward migration of the Subtropical Jet thus permitting the northward surge of the South-west Monsoon.

The monsoonal effect is felt most markedly in Asia because of its great size, and because of its vast transverse relief barrier. To a lesser extent Australia, North America, and the coasts of western South America and West Africa experience the same reversal.

In recent years marked changes appear to have occurred in monsoonal patterns.

'Air Circulation Wheels'

It has been planetary pressure belts are in fact replaced by 'cells' of pressure, both highs and lows, fluctuating with the seasons. From the Azorean and Hawaiian highs, air flows outward in a clockwise direction, so moving on their equatorward sides into the Trade Wind belt, on their poleward sides into the Westerly Wind system. Similarly, from the Subtropical highs in the South Pacific and Atlantic air moves outward in an anti-clockwise direction.

LOCAL PRESSURE AND WIND SYSTEMS IN MID-LATITUDES

In the North Atlantic and North Pacific, land and sea are inter-mingled in a complicated manner. Here, and particularly in the case of the North Atlantic, are busy oceans, constantly crossed by shipping and air-liners, and flanked by technically advanced countries. In this zone of fluctuating pressure systems and winds, and of changeable weather, meteorology has made most advances. The local pressure systems which so markedly affect these latitudes are described as for the northern hemisphere, though similar systems affect the southern hemisphere.

The main problem of forecasting weather is concerned with the prediction of future developments and changes in the local distribution of pressure, and therefore of the air-masses concerned, of differing temperature and humidity, and moving in various directions. The synoptic chart, the chief tool of the forecaster, reveals that the isobars take the form of closed curves, sometimes with prominent bulges.

Just as in world pressure distribution the basic distinction is between areas of relatively high and low pressure, so on a smaller scale a system of closed isobars may enclose either high pressure, diminishing outward from the centre (an *anticyclone*), or one of low pressure increasing outward (a *depression* or *low*).

These are the fundamental types of pressure distribution, although varieties can be recognized. A *wedge* of high pressure lies between two neighbouring depressions, and a ridge of high pressure is a broader region between two more widely separated depressions; these highs tend to bring very fine, but short-lived weather, the 'borrowed day' that is 'too good to last'.

A *secondary depression* is a small unit on the outskirts of a main depression. It may be just a bulge in the isobars, or it may have a separate centre (with closed isobars) of its own, and it may be more intense (with lower pressure) than the primary depression. Most secondaries seem to travel around the primary one, in the northern hemisphere in an anti-clockwise direction. A *trough* is a narrow elongated area of low pressure between two areas of higher pressure, bringing squally winds, with heavy rain or hail.

A col lies between either two depressions or two anticyclones, but is not definite enough to be either a trough or a wedge. The weather is variable and a col is often a difficult problem for a forecaster. Sometimes the chart reveals a pattern of *straight isobars*, when the resulting weather depends on their orientation; if they run north-south, with decreasing pressure eastward, the air-flow will be from a northerly direction, bringing cold weather and probably showers.

The weather in middle latitudes is dominated by an endless procession of these pressure systems moving generally eastward. Each different system brings a specific sequence of weather, which the forecaster tries to interpret by estimating how each successive system will affect the region with which he is concerned. In a depression it is possible to distinguish lines of separation between the air-masses of which it is comprised. These axe *fronts*, and their definition and recognition is a vital factor for a forecaster, since their passage is marked by distinctive changes of weather.

On the old style of weather-map, isobars, wind-arrows (indicating direction and force), and symbols signifying the weather and amount of cloud were shown at each station, and the recorded temperatures added. On the modern type of weather-map, in addition, the various fronts are marked by distinctive lines, against which the isobars often change direction abruptly.

Anticyclones

An anticyclone tends to move in a slow and some-what irregular manner, its slowness usually meaning that the weather is fairly settled for some time. The associated winds are light and variable, and tend to move outwards from the centre, in a more or less clockwise direction in the northern hemisphere. The anticyclone is the scene of extremely slowly descending air currents.

The weather in an anticyclone tends to be dry, warm and sunny in summer. In winter the anticyclone may be associated either with cloudy overcast conditions (the so-called 'anticyclonic gloom'), or with clear, crisp days and sharp frosts by night. Fogs are common in industrial centres.

There is no adequate theory of why mid-latitude anticyclones form and how they move. The simplest way to regard them is as quiescent air-masses between the more active depressions. Sometimes an anticyclone may remain stationary for a considerable rime, preventing the normal movement of depressions, and forcing them to travel around its margins on less usual courses; this is known as a *blocking high*. This may occur in winter over north-western Europe, especially over Scandinavia, so that depressions are forced over either the Norwegian Sea or south-central Europe; this can cause prolonged cold, dry conditions over the British Isles and western Europe, as in the early months of 1947 and 1963.

Conversely a blocking high formed over north-western Europe in July-August 1968, causing depressions to move continuously over southern Britain and western Europe; this gave a wet cloudy summer to those parts, though north-western England and Scotland, no longer affected by the usual frequent depressions because of this blocking high, had an exceptionally fine summer.

Frontal Depressions

Various theories have been put forward to account for these local low-pressure systems in middle and high latitudes. The most satisfactory explanation, developed originally by Norwegian scientists between 1914 and 1918, may be called the wave theory. This assumes that two adjacent masses, one of cold (polar), the other of warm (tropical) air, are separated by a frontal zone, sometimes as much as 80 km (50 miles) across. The cold air is flowing south-westward, the warm air north-eastward, more or less parallel (Fig. 191). 'Waves' are formed as the warm air-mass bulges into the cold mass, forming a salient, at the most northerly extension of which a low-pressure centre develops. When the pressure at the centre is markedly lower than that at the margin, it is said to be a "deep depression'; if the difference is small, it is a 'shallow depression'.

The bulge or 'bay' of warm air gradually becomes a much more defined 'tongue', the *warm sector* of the depression. Its eastward edge, advancing in the direction of general movement (that is, eastward or north-eastward), is the *warm front*, along which the warm air is rising above the cold air which it is overtaking. As the depression develops, the dense cold air undercuts the salient of warm air from the rear, forcing the warm air upward as a *cold front*. It is now realized that conditions at a cold front vary enormously and are much more complex than was thought. If the warm air-mass is for the most part rising over the cold wedge, it is known as an *anafront*, with heavy rain and appreciable temperature change. If the warm air is descending over the cold wedge, it is a *katafront*, with a backing of the wind followed by clearing and improvement in the weather.

As the depression advances, the passage of each front is associated with a distinctive sequence of weather—of cloud (Fig. 191), rain and temperature change. Ahead of the warm front a broad belt of continuous rain falls from a heavily overcast sky, while the wind backs (changes in an anti-clockwise direction) before the front arrives, and then veers (changes in a clockwise direction) as it arrives. At the cold front there is a marked drop in temperature, rain falls in heavy showers sometimes accompanied by thunder, and the wind freshens from a north or north-westerly direction. Sometimes a cold front may be associated with exceptionally stormy conditions, known as a *line-squall*, moving some distance ahead of the front, a phenomenon especially marked in the Mississippi valley in the U.S.A. This is accompanied by a long cloud-roll, strong squalls of wind sometimes of gale force, violent thunderstorms and a downpour of hail.

Gradually the warm front is overtaken by the cold front, which ultimately lifts the warm sector completely off the surface; this is known as an *occlusion*, the centre line of which is an *occluded front*. The cold air in the rear of the depression has now come up against the cold air against which the warm front was originally formed. If the overtaking cold air is colder than the air in front, it is a cold occlusion; if it is not as cold, it is a *warm occlusion*. After occlusion, the depression tends to 'fill up', that is, die away; most depressions arrive over the British Isles in an occluded state.

Such is the life-history of a depression, passing through:

(i) An incipient stage, when a wave is just forming,

(ii) An open stage, when a distinct warm front and sector have developed,

(iii) An occluded stage, and

(iv) A dissolving stage when the depression fills up. As it moves eastward over the British Isles, the weather at any place depends on where this is situated relative to the various fronts. The analysis of large numbers of synoptic charts shows that depressions seem to move more or less parallel to the isobars in the warm sector, and this 'rule of thumb' is one which helps the meteorologist in predicting the future path of the depression.

In any depression there may be more than two fronts. Thus within the cold sector of a depression a temperature discontinuity may form between Polar air which has been incorporated in the depression and accordingly modified, and fresh in bursts of colder Polar air. Such a *secondary front* may complicate greatly the sequence of weather experienced in an 'ideal' depression, and also the problems of a forecaster. Canadian meteorologists believe that depressions over North America frequently have a dual frontal structure involving three or four air-masses, and it seems that frontal waves generally develop in 'families'.

Thermal Depressions

Intense conditions of low pressure may be caused by the concentrated day-time heating of masses of air in summer. On a large scale this is the cause of the low-pressure conditions which provide a focus for in drawn monsoonal winds. On a small scale, thermal depressions form over the hot deserts, resulting in hot, dry 'convection winds'. In middle latitudes they may be associated with thunderstorms; small thermal depressions may move northward across the English Channel, bringing short-lived heavy rain to southern England.

Lee depressions A special type of depression seems to form in the lee of a mountain range, probably the result of a dynamically created eddy as an air-stream crosses it. These depressions occur in the western Mediterranean basin to the south of the Pyrenees-Cevennes-Alpine barrier, along the eastern flanks of the Canadian Rockies, and to the east of the New Zealand Alps in the South Island. This phenomenon is particularly important in the Mediterranean Sea; indeed, it has been estimated that 60 per cent of the depressions affecting that area originate in this way. A PmK air-stream, associated with a cold front, pours southward through the Rhone valley (in its most pronounced form it is known as the *mistral* and bursts into the Mediterranean. The general fall of pressure with the cold front is intensified to form a low-pressure system over the Gulf of Lions, which may then move off south-eastward through the Mediterranean.

Tornadoes

More intense and localized low-pressure whirls may occur over both land and sea. The land storms, or tornadoes, are common in the Mississippi Basin, averaging about 600 a year. The winds are usually anti-clockwise in the northern hemisphere and of immense velocity; it has been estimated that they may attain 800 km (500 miles) per hour, but no measured records can possibly be taken, for they are the most destructive storms in the world, moving across country as a dark twisting funnel-cloud.

They are associated with troughs of low pressure, where cold air from the north and warm damp air from the Gulf of Mexico come into contact along a front. Local heating also causes an uprush of powerful convection currents, like a vortex. They are short-lived, rarely existing for more than an hour, and are usually only two or three hundred metres across, or even less, moving across country at about 65 km ph (40 mph).

Their destructive effects are fortunately limited in area, but the centre of the whirl may cause immense chaos. Not only are buildings destroyed by the terrific winds, but the excessively low pressure at the centre causes buildings to collapse outward because pressure within them is so much greater than that outside. Widespread devastation and loss of life was caused by a series of tornadoes in the Midwestern States of the U.S.A. in April 1965; here they are popularly called 'twisters'.

Occasional tornadoes are experienced in higher latitudes, and there are a few records of their occurrence in Britain. In 21 May 1950, for example, a small tornado which had formed on a line-squall overturned a double-decker bus at Ely, and on 8 December 1954 another damaged Gunnersbury tube-station. The worst ever in Britain was probably in 1703, when 8000 people were killed. These 'temperate tornadoes' are fortunately rare.

Water-spouts

An intense pressure system similar to a tornado, but over the sea, results in a water-spout. From the low-lying base of a cumulonimbus cloud, a whirling cone of dark grey cloud projects downward, gradually elongating until it touches the surface of the water.

The rapidly whirling water-drops are derived both from condensation because lowering of pressure in the centre of the vortex causes cooling, and also from water picked up from the intensely agitated surface of the sea.

TROPICAL LOW-PRESSURE SYSTEMS

So far attention has been concentrated on low-pressure systems in middle and high latitudes. In the Tropics they are also experienced, and they may be divided into two groups: weak and intense.

Weak Tropical Low-pressure Systems

These weak shallow lows are common near the Equator and in the more humid parts of tropical latitudes, especially in the area of the Intertropical Convergence Zone. Their movement is slow and vague, though generally westward, and winds are light, but they are responsible for long spells of rainfall, as a result of their humid unstable characteristics.

There are several theories concerning their formation, none wholly satisfactory; one envisages their formation on the boundary between two air-streams moving in different directions, another postulates the formation of the centre at the meeting-point of three air-streams, a third the convergence of air to the rear of a wave-trough in the North-east or South-east Trades. Certainly there must be conditions of high temperature and humidity.

Intense Tropical Low-Pressure Systems

Small intense low-pressure systems, with a diameter of between 80 and 400 km (50-250 miles), originate in warm oceans within tropical latitudes, mostly between 6° and 20° latitude on either side of the Equator. These appear to develop from a local heat source (or 'warm core'), causing the formation of a wave disturbance. This process, known as cyclogenesis, is indicated by the development of a whorl of towering cumulus clouds. The atmospheric conditions leading to the formation of such a vortex are not yet fully understood, but satellites can now be used to detect incipient storms by their developing cloud patterns. They occur over all the oceans except in the South Atlantic, and are known as hurricanes in the West Indies and the Gulf of Mexico, as typhoons in the China Seas, as cyclones in the Indian Ocean and as willy-willies off the coast of north-western Australia.

The barometric gradient in these lows is extremely steep; indeed, as the storm approaches the pressure may drop by as much as 40 millibars in a few hours, and 950 mb is commonly recorded. The centre of the storm (or the 'eye', a small region about 20 km (12 miles) across) is an area of calm or at most of light, variable winds, but round it may whirl winds of terrific force, sometimes as much as 270 km (170 miles) per hour. Torrential rain, associated with thunderstorms, also occurs. These tropical storms are a menace to shipping, although when warned by radio a ship may sometimes be able to steam out of the probable path, since they move at a rate of only about 16-24 km ph (10-15 mph).

All hurricanes are anxiously tracked by America's National Hurricane Center at Miami, Florida, and their causes and effects are studied at the National Hurricane Research Laboratory. Probably the worst hurricane in living memory was code-named Camille, which swept across the Gulf states in August 1969, causing immense destruction and loss of life.

Occasionally hurricanes develop in high latitudes. An example occurred on the night of 14-15 January 1968, when a hurricane (code-name *Low* Q) crossed Scotland, leaving a trail of damage across Glasgow, killing 20 people and damaging 30,000 houses. Hurricanes from the Caribbean area may penetrate as far north as New York, occasionally into New England and Canada.

SPECIAL WINDS

Named winds In many parts of the world winds of regular or periodic occurrence are sufficiently distinctive, as a result of their characteristics of temperature and humidity, to be given proper names.

Depression Winds

The position and relief of certain regions gives distinctive characters to winds associated with depressions. A moving depression involves air-masses originating both on its poleward and equatorward side, and therefore both cold and warm winds may result. A depression moving eastward through the Mediterranean Sea often results in warm, dry winds from the

Sahara in the warm sector. The *sirocco* in Italy and North Africa, the *leveche* in Spain, and the *khamsin* in Egypt are examples. The *gibli* in Tunisia is like the hot blast which comes from an opened oven-door. They are often hot, dusty and excessively dry, and may do much damage to the blossoms of the vine, leaving them as if frost-nipped. Sometimes they are humid, as a result of crossing the sea, and the sticky heat is even more unpleasant. In other parts of the world the *brickfielders* of Victoria in Australia, dusty winds with temperatures exceeding 38°C (100°F), and the humid *zonda* of the Argentine, are comparable in character.

By contrast, the polar air-masses involved in depressions may cause strong, cold winds; 'bursting' into normally milder areas, they are sometimes called 'polar outbreaks'. In the Mediterranean the *bora* and the *mistral* blow from the 'winter high' over Europe. These bitterly cold winds are felt especially when they blow down a 'funnel' between uplands—the mistral down the Rhone valley, the bora south-westward into the Adriatic. Similar to these are the *southerly-burster* in New South Wales, the *pampero* in the Argentine, the *friagem* or surazo in Brazil, the *norther* in Texas, and the *norte* and *papagayo* in Mexico.

Descending Winds

When an air current crosses a mountain range it is warmed adiabatically the best-known of these winds are the *foehn* in the Alps and the *chinook* in the Rockies. The *foehn* blows when a depression lies to the north of the main ranges of the Alps and moist air is drawn northward over the crest. It is now believed that the *foehn* is in part caused by air-mass turbulence on the leeward side of a mountain range, so that eddies force the air to descend. It is then warmed by adiabatic compression.

The foehn descends the northern slopes as a dry, hot wind, following relief lines such as the valleys of the upper Rhine, the Aar, the Reuss, and the Rhone from Martigny to Lake Geneva. It melts the snow rapidly, often causing widespread avalanches, and temperature can rise 10°C in a few hours. It is so dry that the danger of fire is serious, and in some Alpine villages smoking out-of-doors is prohibited during foehn weather. Similarly, the chinook blows from the west over the Rockies, with even more drastic effects; it can raise the temperature by 15°C or more in less than an hour, and a rise of 15°C in three minutes has been recorded. The effects are the same as those of the *foehn*; the Indian word '*chinook*' means 'snow-eater'. While it can do much good in clearing the land of snow, sporadic periods of the *chinook* early in the year can do much harm, for trees and plants begin a premature budding and animals begin to shed their winter coats, so that a renewal of cold conditions can be disastrous.

Other descending winds are the *samun* in Iran, the *nor-wester* on the eastern side of the South Island of New Zealand, the Berg winds blowing down from the plateau of South Africa towards the coast, and the *Santa Ana* in California.

Convection winds Certain desert winds are the result of intense heating and the formation of ring vortex thermals, with convergent rotating air. On a local scale are the *dust-devils* of the Sahara, on a large scale the swirling, burning, sand-laden *simoom* of the Sahara and the *karabwan* of the Tarim Basin in central Asia.

Land and Sea Breezes

Land and sea breezes are local winds on a diurnal rather than on a seasonal basis. They result when differential heating takes place within a relatively short distance, a condition which occurs most commonly near the coast. During a summer day heating occurs over the land, forming local low-pressure areas into which moves cool air from the sea, although it does not penetrate far inland. This sets in during the afternoon, when heating has reached its maximum. At night the land cools more rapidly than the sea, forming slightly higher pressure. The cooled heavier air sinks down-hill, and moves out to sea as a land breeze during the night.

These winds are most marked in normally calm areas, for on the coasts of Trade-wind islands, for example, their effect is masked by the dominant winds. On the islands and along the coasts near the Equator, however, the sea breeze brings a welcome relief from the stagnant humid heat. In some islands these winds are so regular that fishing-boats go out at night with the land breeze and return the next afternoon with the sea breeze.

Mountain and Valley Winds

Mountain winds (sometimes called *anabatic* winds) blow up a valley during the day, while valley winds (or *katabatic* winds) blow down a valley during the night, where the general circulation of winds is not strong enough to obscure these local influences.

The down-valley winds are fairly easy to explain. The upper slopes are at night chilled rapidly by radiation, and the cool, dense air flows down-hill by gravity, following the valley since this is an obvious line of least resistance. A very pronounced form of this wind is where cold air blows off an ice-cap, over which chilling is intense. This can be felt in Greenland and the Arctic islands, and in Ecuador the *nevados* blows down from the Andean snowfields into the high valleys.

The mountain or up-valley winds are less easy to understand. They blow during the daytime, and are the result of differential heating. A partial explanation is that the air near the mountain slopes is heated by conduction to a greater extent than air at the same level above the valley floors. This causes convectional rising of air from above the slopes, hence air moves up from the valleys to take its place.

The Effect of Relief Barriers

In recent years, much attention has been paid to airflow across transverse relief barriers. One obvious result is the production of relief rainfall on the wind- ward side of the barrier. More complex are the aerodynamic results on the leeward side. When a light wind crosses the barrier, the effect is limited to simple laminar streaming. With a stronger wind, a standing eddy forms, and with still stronger winds *tee-waves* are formed which may develop into complicated turbulence, what is sometimes termed *rotor-streaming*. This may produce striking cloud forms. For example, an easterly or north-easterly wind may blow strongly and gustily down the Crossfell escarpment in the northern Pennines into the Eden valley to the west. This results in the lenticular *helm-cloud* resting on the ridge, a type of banner-cloud, while a few km to the west is a parallel line of cloud known as the *helm-bar*, at the crest of the standing lee-wave.

RECENT CHANGES IN ATMOSPHERIC CIRCULATION

During the last few years some significant, though unexplained, changes in the pattern of atmospheric circulation appear to have occurred, causing in broad terms a southward displacement of climatic zones in the northern hemisphere. It has been suggested by

H.H. Lamb that this results from the dominance of unusually high pressure conditions in the Arctic, not just a short-term cyclical change. Easterly moving low pressure systems of mid-latitudes are now penetrating subtropical regions, while the circulation patterns responsible for the monsoons have also been displaced south. The immediate results include a marked increase in rainfall over North Africa, the Mediterranean and the Middle East, a still more marked and disastrous decline in precipitation over the savanna lands to the south of the Sahara and a partial, in places total, failure of the monsoon rains in north-western India and Pakistan for several years. It has been estimated that the summer monsoonal rains have been reduced by half over large areas of the Indian subcontinent since 1957.

JET STREAM

1. Meaning

The strong and rapidly moving circumpolar westerly air circulation in a narrow belt of a few hundred kilometers width in the upper limit of troposphere is called *jet stream*. The circulation of westerly jet stream is confined between poles and 20° latitudes in both the hemispheres at the height of 7.5-14 km. According to World Health Organization (WHO), 'a strong narrow current concentrated along a quasi-horizontal axis in the upper troposphere or in the stratosphere characterised by strong vertical and lateral wind shear and featuring one or more velocity maxima is called jet stream'. In fact, jet stream was discovered during second world war when American jet bomber fighter planes while flying towards Japan (from east to west) found obstructions of an air circulation which was moving in opposite direction (west to east) resulting into marked reduction in the velocity of jet fighter planes, these planes registered marked increase in their velocity while they used to return to their bases (west to cast). After careful study of this phenomena, it was found that there was a strong upper air circulation from west to east in the upper portion of troposphere which presented obstruction in the free movement of jet fighter planes. Based on this fact, westerly strong meandering upper air circulation was called as jet stream.

Properties

The jet streams are characterized by the following properties.

(1) The circulation of jet streams is from west to east in a narrow belt of a few hundred kilometers width at the height of 7.5-14 kin in the upper troposphere.

(2) On an average, jet steams measure thousands of kilometers in length, a few hundred kilometers in width and a few kilometers (2-4 km) depth.

(3) Generally, their circulation is observed between poles and 20° latitudes in both the hemispheres. These are also called circum-polar whirl because these move around the poles in both the hemispheres.

(4) The vertical wind shear of jet streams is 5-10m/second (18-36 km/hour), meaning thereby the wind velocity above or below jet stream decreases by 18-36 km/hour. Lateral wind shear is 5m/second (18km/hour). The minimum velocity of jet stream is 30m/sccond (108 ton/hour).

(5) Their circulation path (trajectory) is wavy and meandering (Fig. 3.20).

(6) There is seasonal change in the wind velocity in jet streams wherein these become strong during winter season and the wind velocity becomes twice the velocity during summer season. Maximum wind velocity is 480 km (per hour).

(7) The extent of jet streams narrows down during summer season because of their northward shifting while these extend upto 20° latitudes during winter season.

3. Types of Jet Streams

On the basis of locational aspect, jet streams are divided into 5 types :

(1) *Polar front Jet streams* are formed above the convergence zone (40-60 lats.) of the surface polar cold air mass and tropical warm air mass. The thermal gradient is steepened because of convergence of two contrasting air masses. These move in easterly direction but are irregular.

(2) *Subtropical westerly Jet streams* move in the upper troposphere to the north of subtropical surface high pressure belt (at the poleward limit of the Hadley cell in both the hemispheres) *i.e.*, above 30°-350 latitudes. Their circulation is from west to east in more regular manner than the polar front jet streams.

(3) *Tropical easterly jet streams* develop in the upper troposphere above surface easterly trade winds over India and Africa during summer season due to intense heating of Tibetan plateau and play important role in the mechanism of Indian monsoon.

(4) *Polar night Jet streams*, also known as stratospheric subpolar jet streams, develop in winter season due to steep temperature gradient in the stratosphere around the poles at the height of 30 km. These jet streams become very strong westerly circulation with high wind velocity during winters but their velocity decreases during summers and the direction becomes easterly.

(5) *Local jet streams* are formed locally due to local thermal and dynamic conditions and have limited local importance.

4. Index Cycles of Jet Streams

The genesis of jet streams is related to temperature gradient from equator towards the poles, surface high pressure at the poles and genesis of circumpolar whirl above the poles caused by topospheric low pressure. It may be pointed out that surface high pressure is intensified over the surface of arctic region due to subsidence of cooled heavy air during winter season in the northern hemisphere. On the other hand, upper air low pressure develops in the upper troposphere above the high pressure of ground surface of the arctic region. Due to this phenomenon a cyclonic system (west to east) of air circulation in the form of a whirl develops around upper tropospheric low pressure.

The general direction of this circulation is from west to east. The equatorward meandering part of this upper air circulation is called jet stream. The upper air arctic whirl becomes very strong during winter season in the northern hemisphere resulting into maximum south-ward of jet stream upto 20°N latitude. There are changes in the position of extent of jet stream from poles towards equator.

The wavy (meandering) jet stream is called *Rossby waves*. The period of transformation of straight path of jet stream to wavy or meandering path is called *index cycle* which is completed in four successive stages (Fig. 3.20).

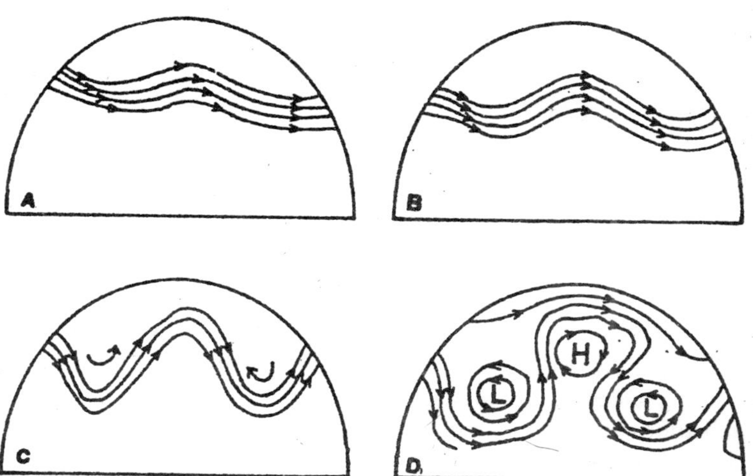

Fig. 3.20 : Index cycle of jet streams.

First Stage

The position of jet streams is near the poles (Fig. 3.20A) and is separated by polar cold air mass in the north and warm westerlies in the south (in the northern hemisphere). The circulation of jet stream is almost in straight path from west to east because Rossby wave is not developed by this time. There is steep pressure gradient across this strong upper air westerly circulation which generates high zonal index.

Second Stage

Gradually, the straight path of jet stream is transformed into wavy path with march of time (Fig. 3.20B). This process initiates the beginning of the development of Rossby waves. With the march of time the amplitude of jet streams increases and gradually they extend towards the equator. The pressure gradient is north-south.

Third Stage

It is characterized by fully developed meandering course of jet stream (Fig. 3.20C), with the result they are positioned near the equator (20° latitude). It may be noted that the pressure gradient in the first two stages is from north to south but in this stage it becomes east-west; There is displacement of polar cold air masses towards equator and of tropical warm air masses towards the poles.

Fourth Stage

It is characterized by cutting off meanders of jet stream from main path due to their more and more meridional circulation (*i.e.*, from north to south) resulting into their circulation in independent circular pattern (Fig. 3.20D) in the form of cyclonic and anticyclonic circulation.

Thus, there develop several cellular circulation patterns which follow cyclonic pattern to the south (L in Fig. 3.20) and anticyclonic pattern in the north (H in Fig. 3.20D). Such cut off low (cyclonic) or cut off high (anticyclonic) air circulation patterns obstruct west to east flow of jet streams.

5. Significance Of Jet Stream

Though, jet streams have not been properly studied as yet but they are supposed to have immense influence on local and regional weather conditions as follows:

(1) There is close relationship between the intensity of mid-latitude (temperate) cyclones and jet streams. These cyclones become very strong and stormy when the upper air tropospheric jet streams are positioned above temperate cyclones of ground surface and yield more precipitation than normal.

(2) There are fluctuations in the local weather conditions due to changes in the form and nature of ground surface cyclones and anticyclones caused by upper air jet streams.

(3) Jet streams cause horizontal convergence and divergence in the upper troposphere. The upper air convergence forms upper air anticyclones while upper air cyclones are developed due to upper air divergence.

(4) The vertical circulation of air in jet streams occurs in two ways *e.g.* cyclonic pattern is characterized by upward vertical air movement while there is downward vertical air movement in anticyclonic pattern of air circulation. This vertical air circulation causes rapid rate of mixing of air between troposphere and stratosphere, which helps in the transport of anthropogenic pollutants from troposphere to stratosphere. For example, the transport of ozone depleting chlorofluorocarbon substances into stratosphere causes global warming.

(5) The monsoon of South Asia is largely affected and controlled by jet streams.

WALKAR CIRCULATION AND EL-NINO SOUTHERN OSCILLATION (ENSO)

Certain variations are found from the atmospheric general circulation patterns *e.g.* surface trades, westerlies and polar winds circulation and tricellular meridional circulation.

Circulation of local and seasonal (monsoon) winds may be cited example of such deviations. East-west zonal circulation of tropical winds is an important variant from general atmospheric circulation.

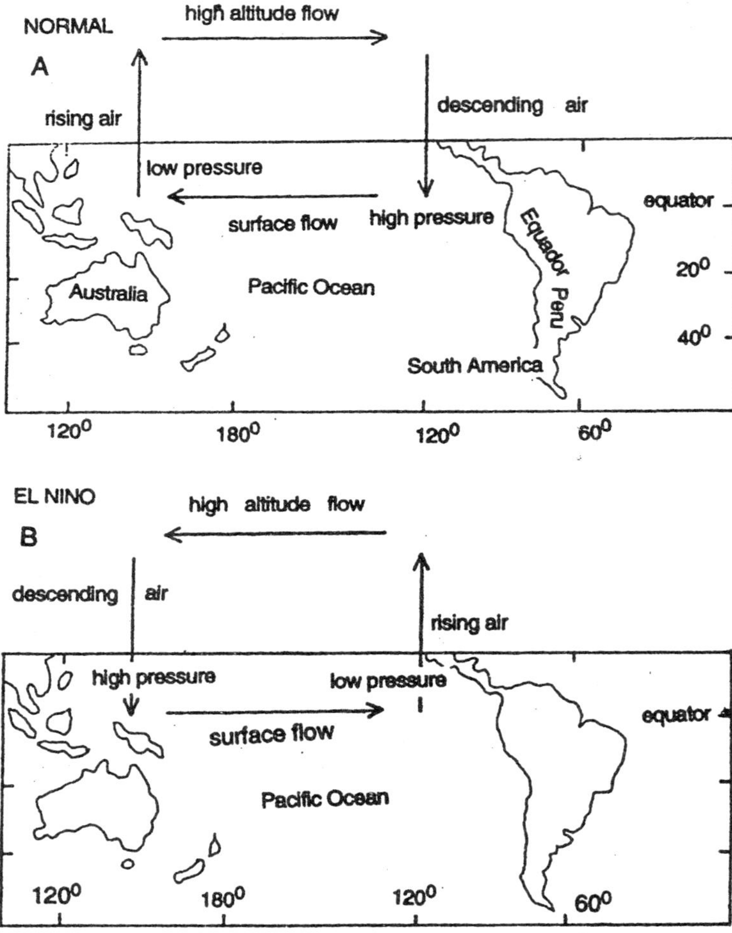

Fig. 3.21 : Southern oscillation. Walkar circulation and El Nino.

This typical east-west circulation of tropical wind is called *Walkar circulation* named after famous scientist G.T. Walkar.

In fact, Walkar circulation is a convective cell of air circulation, which is formed due to the development of pressure gradient from east to west in the equatorial Pacific ocean. After two-three years this general condition of east-west pressure gradient is reversed *i.e.,* pressure gradient becomes from west to east (Fig. 3.21B). Thus, there are oscillations in pressure gradient and air circulation after the intervals of 2-3 years. Walkar called such oscillation as southern oscillation.

Walkar circulation and southern oscillations are driven by the sea surface pressure gradient from the equatorial eastern Pacific ocean (near the western coastal areas of South America, to the equatorial western Pacific ocean (near S-E Asian coasts). In normal conditions high pressure develops on the sea surface of the equatorial east Pacific ocean and the western coastal lands of south America (Fig. 3.19A) due to subsidence of air from above and upwelling of cold oceanic water. On the other hand, low pressure is formed in the equatorial western Pacific ocean due to rise of air from the warm sea surface. This pressure gradient from east to west generates east-west circulation of trade winds on the surface while there is reverse upper air circulation *i.e.,* from west to east, (Fig. 3.21A) which completes a convective cell. This east-west air circulation drives the ocean water mass from the western coast of South America towards the west. This phenomenon facilitates upwelling of cold sea water near the coasts of Peru and Equador resulting in further cooling of air, high air pressure, atmospheric stability and dry weather condition. Contrary to this, east-west air circulation becomes warm north-east trades in the equatorial west Pacific ocean where it, after being heated, rises upward, becomes unstable and causes precipitation. After rising to certain height it turns eastward and descends in the equatorial eastern Pacific ocean to complete the convective cell (Fig. 3.21A). This is now evident that tropical eastern and western Pacific is characterized by dry and wet weather conditions respectively.

By October-November the low air pressure of the tropical western Pacific is shifted to the tropical eastern Pacific causing weakening of trade winds. This reversal in pressure condition facilitates the return of warm sea water which was driven from the coasts of South America westward, towards the tropical east Pacific. Consequently, low air pressure is formed in the south-east Pacific mainly off the coasts of South America (Equador and Peru), upwelling of cold seawater is stopped, warm air-rises upward and becomes unstable and ultimately yields rainfall after condensation. It is evident that the general normal condition (Fig. 3.21A) has got reversed (3.21B). This event is called El Nino phenomenon. The rising air in the east Pacific cools above and turns westward in the troposphere and ultimately descends in the tropical west Pacific giving birth to high pressure which drives warm air towards the coasts of South America. Thus, again a complete convective cell is formed. Such condition is called El Nino-Southern Oscillation Event (ENSO Event). In fact, changes in the positions of air pressure in the tropical eastern and western Pacific are called southern oscillations. During El Nino event Walkar circulation is weakened due to the development of equatorial westerlies on sea surface (Fig. 3.21B) but Hadley circulation is activated. This phenomenon again activates trade winds which again drive sea-water of the tropical eastern Pacific west-ward resulting in the upwelling of cold water from below, weakening of El Nino event and re-establishment of normal condition (Fig. 3.21A).

INSOLATION AND HEAT BUDGET

MEANING AND DEFINITION

The earth receives heat energy from three basic sources viz. (i) solar radiation, (ii) gravity, and (iii) endogenetic forces coming from within the earth but the solar radiation is the most significant source of terrestrial heat energy. "The radiant energy received from the sun, transmitted in a form analogous to shortwaves (1/250 to 1/6700mm in length), and travelling at the rate of 1,86,000 miles a second, is called solar radiation or insolation' (G.T. Trewartha). The sun is a great engine that drives winds on the earth's surface, ocean currents, exogenetic or denudational processes and ultimately sustains life in the biosphere. Solar energy received from the sun through solar radiation heats the earth's surface and the atmosphere and thus is responsible for the movement of air and currents through changes in pressure gradients, drives the hydrological cycle through evaporation and precipitation which in turn helps in the cycling and recycling of nutrients and chemical elements in the biosphere through the broader cyclic pathways collectively known as 'geobiochemical cycles', helps the plants to prepare their food through the process of photosynthesis' which infact, changes solar energy into chemical energy which is used by plants, animals and man, through different trophic levels of food chains and food webs.

MECHANISM OF SOLAR RADIATION

The source of energy of the sun is its interior wherein the hydrogen is converted into helium due to enormous confining pressure and very high temperature under the process of nuclear fusion which generates huge quantity of heat. This heat is transported to the outer surface of the sun through convection and conduction from below. It may be pointed out that the rate of generation of heat inside the sun is more or less constant and hence the radiation of energy from the outer surface of the sun (called as *photosphere*) is also more or less constant. Thus, the amount of solar radiation or the solar energy received on a unit area of the surface facing the sun at the average distance between the sun and the earth is also more or less constant and is called as solar constant. Thus, it is obvious that the solar constant refers to the rate of radiation from the sun which is of the value of 2 gram calories per square centimetre per minute (2 cal/cm^{2}min). It is also expressed in terms of langley (a unit measure of heat energy, one gram calorie per square centimetre is equal to one langley) as 2 langley per minute (2ly/min).

There are two basic laws which govern the nature and flow of radiation as given below :

(1) *Wien's displacement law* 'states that the wavelength of the radiation is inversely proportional to the absolute temperture of the emitting body.'

(2) *Stefan-Boltzmann law* states 'that flow, or influx, of radiation is proportional to the fourth power of the absolute temperature of the radiating body.'

The surface temperature of the sun is 6000°C or 11000°F. The highly incandescent gas of the sun's surface being heated from below emits bundle of energy called 'photon' which is infact the particle of radiation which has the property of wavelength. Continuous emission of photons from the sun's surface causes continuous bands of radiation having certain wavelength which is considered as short wavelength in relation to the earth's outgoing longwave radiation. The solar energy radiated from the outer surface of the sun in the form of electromagnetic wave is called as *electromagnetic radiation*, which travels at the speed of 3,00,000 km per second (1,86,000 miles per second). The solar energy received at the earth's surface is called *insolation* or *solar radiation*.

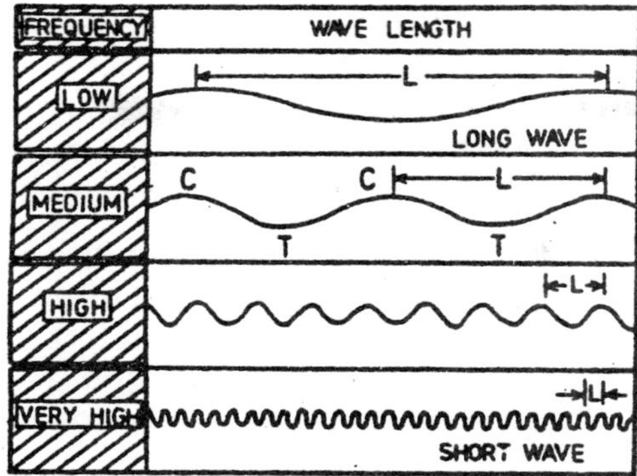

Fig. 4.1: Relationship between wavelength and wave frequency, L = wavelength, C = wave crest, and T = wave trough.

The energy from the sun is emitted in the form of electromagnetic waves which travel outward in radial manner from the sun almost in straight line and take 8 minutes 20 seconds to reach the earth's surface after covering an average distance of 150 million km (93 million miles) between the sun and the earth. The electromagnetic radiation waves are expressed in terms of wavelengths (L). The straight distance between two successive crests or troughs is called wavelength (Fig. 4.1) (L) which is expressed in the length units of metres, centimetres, millimeters, microns etc.

The number of radiation waves (one radiation wave is equal to one wavelength) passing through a certain point per unit time (usually one second) is called *wave frequency* which varies according to the wavelengths of the radiation waves. There is inverse relationship between the wavelength and wave frequency *i.e.,* shorter the wavelength, higher the wave frequency and longer the wavelength, lower the wave frequency (Fig. 4.1). In other words, high wave frequency is associated with short wavelength and low wave frequency is associated with long wavelength. Wave frequency is generally expressed as wave cycles per second. The wave cycles are usually expressed by the unit of measure of hertz. For example, one hertz per second represents one wave cycle meaning thereby only one wavelength passes per second from a fixed point. Hertz is further expressed in kilohertz (1,000 hertz) or megahertz (1,000,000 hertz).

The electromagnetic radiation emitted from the outer surface of the sun consists of four spectra of radiation waves having different wavelengths and wave frequencies.

(1) The first spectrum of the electromagnetic waves includes *gama rays, hard x-rays, soft x-rays* and *ultraviolet rays.* The wavelengths of this spectrum of the shortest wavelengths are expressed in the unit measure of 'angstrom' wherein one *angstrom* is equal to 0.000,000.01 cm or 10^{-8}cm. The following are the wavelengths of the waves of the spectrum of the shortest waves.

Waves	*Wavelength*	*Frequency, of wave-lengths in angstrom in megahertz*
1. Gama Rays	less than 0.03	10^{14}
2. Hard X-Rays	0.03-0.6	10^{13}
3. Soft X-Rays	0.6-100	10^{13}
4. Ultra-Violet Rays	100-4000	10^{10}-10^{9}

(2) The second spectrum of the electromagnetic radiation waves is also called as the spectrum of visible light or rays which includes violet, blue, green, yellow, orange and red rays which carry 41 per cent of the total energy of the solar spectrum of all the electromagnetic radiation waves. The unit of the measure of wavelengths of this spectrum is micron (one micron is equal to 0.0001 cm or 10,000 angstroms). The wave frequency of these different rays ranges between 109 and 108 megahertz per second. The wavelengths of these visible rays are as given below:

Visible Rays	*Wavelengths in microns*
1. Violet rays	0.4 – 0.43
2. Blue rays	0.43 – 0.49
3. Green rays	0.49 – 0.53

4. Yellow rays 0.53 – 0.58
5. Red rays 0.58 – 0.70

(3) Third spectrum of the electromagnetic radiation waves is called as infrared spectrum which consists of infrared waves of the wavelengths ranging from 0.7 micron to 300 microns. The wave frequency ranges between 108 and 106 megahertz per second.

(4) Fourth spectrum of the electromagnetic radiation wave consists of longwaves including *microwaves, radar waves* and *radio waves*. The unit measure of these wavelengths of longwaves is usually centimetre to metre. The wavelengths of microwaves range between 0.03 cm and 1.0 cm. These waves are used to send messages from one place to other distant places. The wavelengths of radar waves vary from 1.0 cm to 100 cm (1 m). The radar system is divided into two sub-systems on the basis of frequency viz. (i) *radio system* and (ii) *television system*. The radar, television and radio waves are divided into 6 categories on the basis of wave frequency and wavelengths.

Frequency	*Wavelengths*
1. Extremely High Frequency (EHF)	0.1–1.0 cm
2. Super High Frequency (SHF)	1.0cm–10.0cm
3. Ultra-High Frequency (UHF)	10.0 cm–100.0 cm
4. Very High Frequency (VHP)	1.0m–10.0m
5. High Frequency (HF)	10m–100m
6. Low Frequency (LF)	more than 100m

DISTRIBUTION OF INSOLATION

On an average, the amount of insolation received at the earth's surface decreases from equator towards the poles but there is temporal variation of insolation received at different latitudes at different times of the year. Table 33.1 depicts the amount of insolation received at the outer boundary of the atmosphere and at the earth's surface at the time of winter solstice (22 December), vernal equinox (21 March), summer solstice (21 June) and autumnal equinox (23 September) as given by Baur and Phillips.

Table 4.1 : Average Amount of Direct Solar Radiation Received at the outer Boundary of the Atmosphere and at the Earth's Surface (in cal/cm^{2}min).

Date	*Latitudes (northern hemisphere)*						
	0-10	*10-20*	*20-30*	*30-40*	*40-50*	*50-60*	*60-90*
	A–Received at the upper limit of the atmosphere						
December, 22	0.549	0.465	0.373	0.274	0.173	0.079	0.006
March, 21	0.619	0.601	0.563	0.509	0.441	0.358	0.211
June, 21	0.579	0.629	0.664	0.684	0.689	0.683	0.703
September, 23	0.610	0.592	0.556	0.503	0.435	0.353	0.208
	B - Received at the earth's surface if cloudiness and turbidity are considered						
December, 22	0.164	0.161	0.134	0.082	0.036	0.013	0.001
March, 21	0.191	0.221	0.206	0.161	0.116	0.096	0.055
June, 21	0.144	0.170	0.216	0.233	0.183	0.159	0.133
September, 23	0.170	0.162	0.201	0.183	0.131	0.079	0.028

Source : Baur and Phillips

It is apparent from Table 4.1 that the amount of solar radiation reaching the outer limit of our atmosphere is significantly more at different latitudes (A in Table 4.1) than the amount of insolation received at the ground surface. This trend reveals the fact that a sizeable portion of incoming solar radiation is lost while passing through the atmosphere due to cloudiness, atmospheric turbidity (scattering), reflection, and absorption (through ozone).

The data of insolation as portrayed in Table 4.1. A further reveal that maximum insolation reaches the outer limit of the atmosphere at north pole at the time of summer solstice while maximum insolation is received at the ground surface between latitudes 30°-40°N on 21st June because of minimum amount of cloudiness due to the presence of subtropical high pressure belt and anticyclonic conditions.

Table 4.2: Amount of insolation received at the earth's surface from equator towards the poles (in percentage).

Latitudes	0	10	20	30	40	50	60	70	80	90
Insolation in per cent	100	99	95	88	79	68	57	47	43	42

Table 4.2 reveals the fact that the total amount of insolation received at the earth's surface decreases from equator towards the poles. The insolation becomes so low at the poles that they receive about 40 per cent of the amount received at the equator. The tropical zone extending between the tropics of Cancer (23.5°N) and Capricorn (23.5°S) receives maximum insolation. Not only this, there is very little variation of insolation during winter and summer seasons because every place experiences overhead sun twice every year. The globe is divided into 3 zones on the basis of the amount of insolation received during the course of a year.

(1) *Low latitude* or *tropical zone* extends between the tropics of Cancer and Capricorn. All places experience overhead sun (sun's rays are vertical) twice during the course of a year due to northward and southward march of the sun. Consequently, every place receives maximum and minimum insolation twice a year. The region receives highest amount of insolation of all other zones and there is little seasonal variation.

(2) *Middle latitude zone* extends between 23.5° and 66° latitudes in both the hemispheres. Within this zone every place receives maximum (at the time of summer solstice—21 June in the northern hemisphere and at the time of winter solstice—22 December in the southern hemisphere) and minimum (at the time of vernal equinox—21 March in the northern hemisphere and at the time of autumnal equinox—23 September in the southern hemisphere) insolation once during the course of a year. Insolation is never absent at any time of the year but seasonal variation increases with increasing latitudes.

(3) *Polar zone* extends between 66° and 90° (poles) latitudes in both the hemispheres. Every place receives maximum

and minimum insolation once during the course of a year but some times insolation becomes zero due to absence of direct solar rays.

FACTORS AFFECTING THE DISTRIBUTION OF INSOLATION

It is apparent from the foregoing discussion that the amount of insolation received at the earth's surface varies significantly (decreases) from equator towards the poles due to certain astronomical and geographical factors viz. :

(i) Angle of the sun's rays,

(ii) Length of day,

(iii) Distance between the sun and the earth,

(iv) Sunspots, and

(v) Effects of the atmosphere.

(1) Angle of the Sun's Rays

The angle between the rays of the sun and the tangent to the surface of the earth at a given place largely determines the amount of insolation to be received at that place. The sun's rays are more or less vertical (maximum angle of 90° between the sun's rays and the tangent to the earth's surface) at the equator and become more and more oblique poleward. In other words, the angle of the sun's rays decreases poleward. As per rule vertical rays bring more insolation than oblique rays. In other words, as the angle of sun's rays decreases poleward, the amount of insolation received also decreases in that direction. The control of the angle of the sun's rays on the amount of insolation may be explained with the following examples :

(i) Vertical rays are spread over minimum area of the earth's surface and they heat the minimum possible area and thus the energy received per unit area increases. On the other hand, oblique rays are spread over larger area of the earth's surface and hence the amount of energy received per unit area decreases. It is, obvious from Fig. 4.2 that A and B bands of the sun's rays are of reinforce width and carry equal amount of solar energy but the area (S) covered by A band is much smaller than the area covered by B band (O) and therefore the amount of insolation received per unit area over S surface is greater than over O surface area (Fig. 4.2)

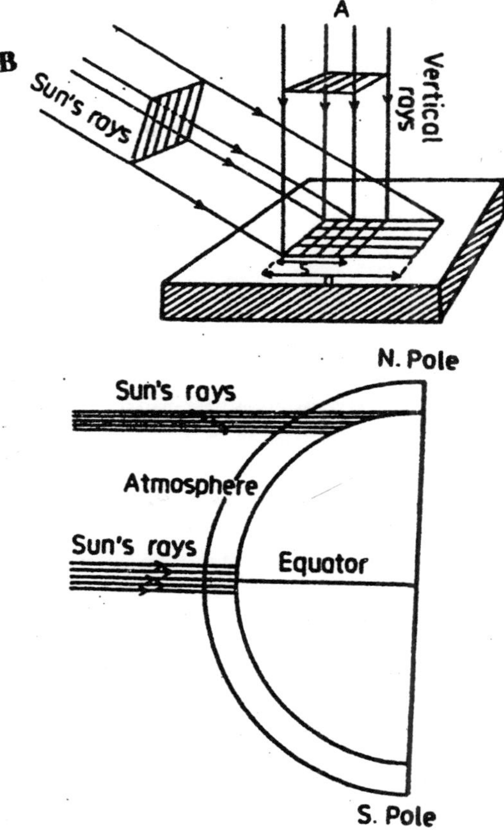

Fig. 4.2 : Effect of the angles of sun's rays on the distribution of insolation.

(ii) Oblique rays have to pass through thicker portion of the atmosphere than the vertical rays. Thus, the oblique rays have to traverse larger distances than the vertical rays. Consequently, the amount of solar energy lost due to reflection, scattering and absorption increases with increasing distance of travel path covered by the sun's rays through the atmosphere (Fig. 4.2). It may be summarized that the oblique rays lose more energy than the vertical rays while passing through the atmosphere.

(2) Length of Day

If all the other conditions are favourable and equal then longer duration of sunshine (or length of day) and shorter duration of night enable the ground surface to receive larger amount of insolation. On the other hand, shorter the duration of sunshine and longer the period of night, the lesser the amount of insolation received at the earth's surface. The length of day varies at all places except at the equator due to inclination of the earth's axis, its parallelism and the earth's rotation and revolution. The length of day is always of 12 hours at the equator because the circle of illumination or light circle always divides the equator into two equal halves. But the length of day increases poleward with northward march of the sun in the northern hemisphere while it decreases in the southern hemisphere at the time of summer solstice (21 June). On the other hand, the length of day increases from the equator poleward in the southern hemisphere but it decreases in the northern hemisphere at the time of winter solstice (22 December) (southward march of the sun). It is important to note that the duration of day becomes of 6 months at the north pole from 21 March to 23 September during the northward migration of the sun while the night is of the duration of 6 months at the south pole during this period. Conversely, the length of day becomes of 6 months at the south pole (23 September to 21 March) during the southward migration of the sun while the night becomes of 6 months at the north pole during this period. Inspite of increasing length of day from the equator towards the north pole during summer solstice and from the equator to the south pole during winter solstice amount of insolation received at the ground surface decreases considerably poleward because of decrease in the angle of sun's rays. Inspite of the longest length of day at the poles insolation becomes minimum because :

(i) the sun's rays become more or less parallel to the ground surface, and

(ii) the ice cover reflects most of the solar radiation. It is apparent that the angle of the sun's rays controls the amount of insolation received more effectively than the length of day. It may be, thus, concluded that the places having longer length of day and vertical sun's rays will certainly receive maximum insolation.

Table 33.3: Maximum length of day at different latitudes

Latitude	0	17	31	41	49	58.5	63.4	66.5	67.4	69.8	78.2	90
Length of day (hours)	12	13	14	15	16	18	20	24	1 month	2 months	4 month	6 months

(3) Distance between the Earth and the Sun

The distance between the sun and the earth changes during course of a year because the earth revolves around the sun in elliptical orbit. The average distance between the sun and the earth is about 93 million miles (149 million kilometres). At the time of perihelion on January 3 the earth is nearest to the sun, say 91.5 million miles (147 million kilometres) away while at the time of aphelion on July 4 it is farthest from the sun, say 94.5 miles (152 million kilometres) away (Fig. 4.3). As per rule, the earth at the time of perihelion, when it is nearest to the sun, should receive maximum insolation while it should receive minimum insolation at the time of aphelion when the earth is at the greatest distance from the sun. In fact, in the month of January, when the earth is nearest to the sun, there is winter season instead of summer season in the northern hemispheres due to low amount of insolation received. On the other hand, in the month of July, when the earth is farthest from the sun, there is summer instead of winter in the northern hemisphere due to high amount of insolation received. It is obvious that factors of the angle of the sun's rays and length of day play more dominant role in the distribution of insolation than the factor of varying distances between the earth and the sun. Of course winters are 7 per cent less severe in January in the northern hemisphere but summer is 7 per cent more intense in the southern hemisphere at the time of perihelion while summer is 7 percent less intense in July in the northern hemisphere but winter is 7 per cent more intense in the southern hemisphere at the time of aphelion.

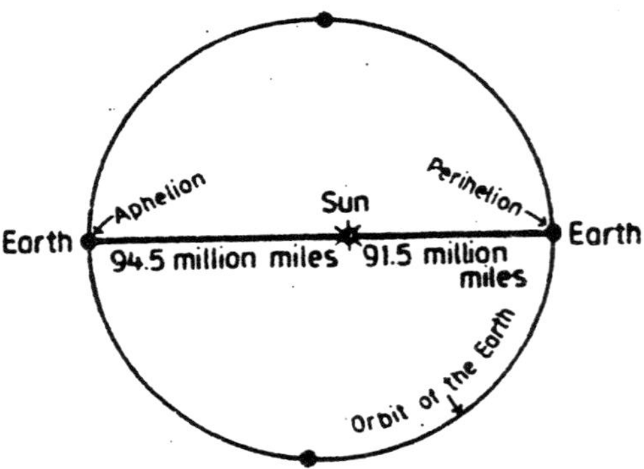

Fig. 4.3 : Relative distance between the sun and the earth.

(4) Sunspots

Sunspots are created in the solar outer surface due to periodic disturbances and explosions. The number of sunspots varies from year to year. The studies have shown that the variation in the number of sunspots is cyclic in nature. In other words, the increase and decrease of number of sunspots is completed in a cycle of 11 years. During every 11th year there is maximum number of sunspots. The energy radiated from the sun increases when the number of sunspots increases and therefore the amount of insolation received at the earth's surface also increases. On the other hand, the amount of insolation received at the earth's surface decreases when the number of sunspots decreases.

(5) Effects of the Atmosphere

The electromagnetic solar radiation or the incoming shortwave solar radiation has to pass through thick layer of the earth's atmosphere and hence it is partly absorbed, partly reflected and partly scattered by the atmosphere and partly transmitted to the earth's surface.

Absorption

If the total amount of energy radiated from the sun towards the earth and its atmosphere (which is 1.2 billionth part of the total energy radiated from the photosphere of the sun) is taken to be 100 per cent, about 14 per cent of this amount is absorbed by the atmospheric gases (*e.g.* by ozone in the stratosphere to larger extent and oxygen and carbon dioxide to very limited extent), water vapour, haze etc. The process of absorption is selective in nature. The shortest wavelengths ranging between 0.02 micron and 0.29 micron are absorbed by oxygen (O_2) and ozone (O_3) gases. Ozone also absorbs ultraviolet rays of the wavelengths varying from 1000 angstroms to 4000 angstroms and thus prevents these ultraviolet radiation waves from reaching the earth's surface. Water vapour absorbs the incoming solar radiation waves of the wavelengths ranging between 0.9 micron and 2.1 microns.

Scattering

Some portion of the incoming electromagnetic solar radiation (23%) is scattered in the atmosphere by dust particles and haze. Six per cent of this scattered energy is sent back to space while 17 per cent reaches the earth's surface. The process of scattering is selective in nature. Scattering becomes possible when the diameter of invisible dust particles suspended in the air and the molecules of the atmospheric gases is shorter than the wave lengths of the solar radiation waves. Blue light of the incoming shorter wavelengths is more scattered than red light. This is the reason that the sky Similarly, the picturesque reddish hue of the sky during sunrise (dawn) and sunset (twilight) is the result of scattering of all the colour spectra except the red and orange because at the time of sunrise and sunset the oblique rays have to pass through the longest path of the atmosphere.

Reflection

The scattering of incoming solar radiation waves by dust particles and molecules of water vapour (clouds) when the diameter of these particles is longer than the wavelengths of incoming solar radiation is called *diffuse reflection* which sends some portion of incoming solar energy back to space while some portion remains in the lower atmosphere. The diffused and scattered solar energy present in the lower atmosphere enables us to see even the dark portion of the moon. One can also see (if not suffering from cataract) even in the pitch darkness of night. Some of the scattered and diffused solar energy reaches the ground surface. Such energy is called as *diffuse blue light* of the sky or diffuse day light. Some portion of incoming solar radiation is reflected back to space by high clouds (27 per cent) and by the ice-covered ground surface (2 per cent).

The portion of incident radiation energy reflected back from a surface is called *albedo*. Various attempts have been made to measure total albedo of the earth (including its atmosphere). Various data derived so far indicate the earth's average albedo fluctuating between 29 per cent and 34 per cent (including the energy reflected through the mechanism of diffuse reflection by dust particles, water molecules etc., (from the cloud surface and from the earth's surface). The albedo of other planets has also been estimated *e.g.* Moon (7%), Mercury (6%), Mars (16%), Venus (76%) and the remaining outer planets (73% to 94%).

It may be pointed out that the processes of absorption, scattering and reflection are not as simple as discussed above rather they are highly complex. Further more, the figures used here to indicate the quantity of solar radiation lost during its passage through the atmosphere by different processes are mere estimates and these vary from the estimates of one scientist to the other.

HEAT BUDGET OF THE EARTH AND THE ATMOSPHERE

'The total solar radiation reaching a horizontal surface on the ground is called global radiation. It comprises the direct shortwave radiation from the sun, plus the diffuse radiation scattered by the atmosphere' (J.E. Hobbs, 1980). On an average, there is supposed to exist heat balance between the amount of solar radiation received by the earth's surface and its atmosphere and the amount of heat lost by the outgoing terrestrial longwave radiation from the earth's surface and loss of heat from the atmosphere. It may be pointed out that the solar energy received at the earth's surface is convened into heat energy which heats the outer surface of the earth. Thus, the earth after being heated also radiates energy in the form of longwave radiation. The radiation from the sun towards the earth and from the earth towards the atmosphere and the space is called *incoming shortwave solar radiation* (radiation from the sun) and *outgoing longwave terrestrial radiation* (from the earth) respectively.

The estimated data of G.T. Trewarha are being used to explain the heat budget of the earth and its atmosphere.

(1) Incoming Shortwave Solar Radiation and the Heat Budget of the earth and the Atmosphere

The earth receives most of its energy from the sun through shortwave solar radiation. The solar energy radiated towards the earth's surface (1/2 billionth part of the total energy radiated from the outer surface-*photosphere* of the sun which is equivalent to 23 trillion horse power) is taken as 100 per cent or 100 units. Out of the total incoming solar radiation entering the earth's atmosphere 35 per cent is sent back to space through scattering by dust particles (6%), reflection from the clouds (27%) and from the ground surface (2%), 51 per cent is received by the earth's surface (received as direct radiation), and 14 per cent is absorbed by the atmospheric gases (ozone, oxygen etc.) and water vapour in different vertical zones of the atmosphere. The 51 per cent solar energy received by the earth comprises 34 per cent as direct solar radiation and 17 per cent as *diffuse day light*. The heat budget of the atmosphere comprises 48 per cent of solar radiation wherein 14 per cent is received through absorption of the shortwave incoming solar radiation and 34 per cent is received from the outgoing longwave terrestrial radiation.

(i) Incoming shortwave solar radiation = 100%

(ii) Amount lost to space through scattering, and reflection

(a) reflected by the clouds = 27%

(b) reflected by the ground = 2% } 35%

(c) scattered energy lost to space = 6%

Remaining solar energy = 65%

(A) **Heat budget of the earth**

(i) Received through direction radiation	= 34%
(ii) Received as diffuse day light	= 17%
Total	= 51%

(B) **Heat budget of the atmosphere**

(i) Absorption of incoming solar radiation	= 14%
(ii) Received from outgoing terrestrial radiation	= 34%
Total	= 48%

(2) Outgoing Longwave Terrestrial Radiation and Heat Balance

After receiving energy from the sun (Fig. 4.4) the earth also radiates energy out of its surface into the atmosphere through longwaves (Fig. 4.4). The terrestrial radiation is also called 'effective radiation' because it helps in heating the lower portion of the atmosphere. Twenty three per cent energy (out of 51% energy which the earth has gained from the sun) is lost through direct longwave outgoing terrestrial radiation out of which 6 per cent is absorbed by the atmosphere and 17 per cent goes directly to the space. About 9 per cent of the terrestrial energy is spent in convection and turbulence and 19 per cent is spent through evaporation which is added to the atmosphere as latent heat of condensation. Thus, the total energy received by the atmosphere from the sun (14%) and the earth (34%) becomes 48 per cent which is reradiated to the space in one way or the other.

(i) **Terrestrial Heat Balance**

Energy received	*Energy lost*
51%	(i) 23% through radiation
	(ii) 9% through convection and turbulence
	(iii) 19% through evaporation total = 51%

(ii) **Atmospheric Heat Balance**

Energy received	*Energy lost*
(i) 14% through absorption of incoming solar radiation	
(ii) 6% through effective radiation from the earth	48% through radiation into the space
(iii) 9% through convection and turbulence from the earth	
(iv) 19% through evaporation total = 48%	

Energy sent back to space = 35% + 17% (through radiation from the earth) + 48% (through radiation from the atmosphere) = 100%.

It may be pointed out that the mechanism of solar and terrestrial radiation is not as simple as mentioned above, rather it is highly complex. For example, not all the energy received by the atmosphere from the sun and the earth is reradiated directly to the space rather a sizeable amount of energy received by the atmosphere is counter-radiated to the earth's surface which is again radiated to the space and the atmosphere. Table 4.4 presents a complex radiation balance of the earth and the atmosphere.

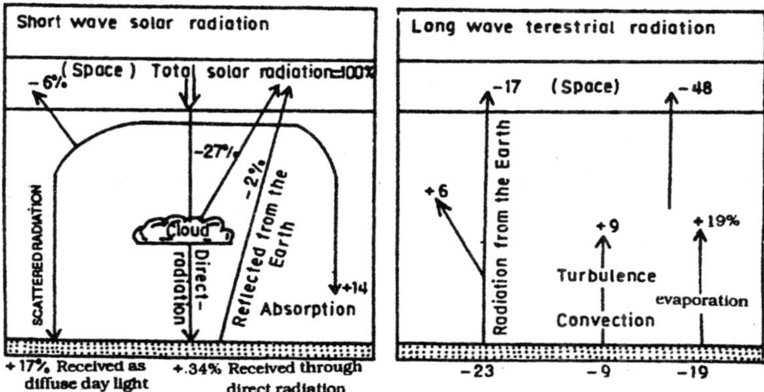

Fig. 4.4: Radiation or heat balance of the earth and the atmosphere.

Table 4.4: Global Heat Budget of the Earth and the Atmosphere

Incoming Solar Radiation	*Percentage*
Total amount of energy at the top of the atmosphere (equal to 263 kilo angles per year)	100
(a) Diffuse reflection to space by scattering	5
(b) Reflection from clouds to space	21
(c) Direct reflection from earth's surface	6
Total reflection lost to space by earth-atmosphere system (earth's albedo)	32
(a) Absorbed by molecules, dust, wa:er vapour, carbon dioxide, clouds etc.	18
(b) Absorbed by earth's surface	50
Total absorbed by earth-atmosphere system	68
Sum of absorption and reflections	100
Outgoing Longwave Terrestrial Radiation	Percentage
(A) Radiation Balance of the Earth's surface	
(a) Loss of energy directly to space	8
(b) Loss of energy to atmosphere	90
Total radiation loss from earth's surface	98
(c) Gain by earth's surface by counter radiation from the atmosphere	77
Net outgoing radiation from earth's surface	21
(B) Radiation Balance of the Atmosphere	
(a) Gain of energy from the radiation from the earth's surface	90
(b) Loss of energy through counter radiation from the earth's surface	77
(a) Net gain of energy from the radiation from the earth's surface	13

(a) Gain of energy from direct shortwave absorption	18
(c) Gain of energy from latent heat transfer	20
(d) Gain of energy from mechanical heat transfer	9
Total Net Gain by the Atmosphere	60
Radiated to Space from the Atmosphere	68
Radiated directly to Space from the Earth's Surface	8
Total radiation from the planet Earth to Space	68

Source of Data: W.D. Sellers (1965)

NET RADIATION AND LATITUDINAL HEAT BALANCE

The difference between all incoming solar energy and all outgoing terrestrial energy by both shortwave and longwave radiation is called net radiation. It is apparent from the aforesaid discussion that the net radiation from the whole globe is at least theoretically zero but this is far from truth if we look at the regional distribution of insolation. There are some places where the receipt of solar energy is more than the energy lost because the solar energy comes at faster rate than the terrestrial energy goes out. Similarly, in some areas the loss of energy through outgoing terrestrial radiation is faster than the gain of incoming solar radiation. This mechanism results in the development of areas of energy surplus (where incoming solar radiation exceeds outgoing terrestrial radiation and energy deficit (where outgoing terrestrial radiation exceeds incoming solar radiation).

Table: 4.5 Annual Meridional Heat Transport

Latitudes (northern hemisphere)	*Heat transport (kilo calories per year × 10^9)*	*Latitudes (northern hemisphere*	*Heat transport (kilocalories per year × 10^9)*
90	0.00	40	3.91
80	0.35	30	3.56
70	1.25	20	
60	2.40	10	2.54
50	3.40	09 (equator)	–0.26

Source: W.D. Sellers, 1965.

The energy surplus and energy deficit areas may be identified and studied at two levels viz. :

(i) *At the earth's surface*, and

(ii) *In the atmosphere*, latitudinal base being common in both the cases,

(iii) The distribution of net radiation at the earth's surface from equator towards the poles shows that :

(a) there is large *energy surplus area* between the zones of 20°N and 20°S where energy gain from the incoming solar radiation is more than the loss of energy through outgoing longwave terrestrial radiation and the annual surplus energy amounts to 100 kilolangleys per year;

(b) net radiation rapidly decreases from the energy surplus areas of low latitudes towards mid-latitudes;

(c) net radiation becomes practically zero near 70° latitude in the northern and southern hemispheres, and

(d) the polar areas are the zones of perennial energy deficit

The latitudinal distribution of net radiation in the atmosphere shows that 'the atmosphere is itself a net loser of radiation at all latitudes' (J.E. Hobbs, 1980). Thus, the atmosphere is the zone of perennial energy deficit because the deficit of energy always exceeds 60 kilolangleys per year.

If the data of net radiation of both, the earth's surface and the atmosphere, are combined together, the net radiation value for the combined '*earth's surface-atmosphere system*' may be calculated. Based on the combined data the following energy zones are identified :

(a) large region of surplus radiation extending between 40°N and 30°S latitudes,

(b) northern high-latitudes of deficit radiation and

(c) southern high-latitudes of deficit radiation (A.N. Strahler, 1978).

It means that "there must exist a two-way heat transfer; from the earth's surface to the atmosphere and from the equator to the poles" (J.E. Hobbs, 1980). This can be achieved if heat is transported from the earth's surface to the atmosphere and from the tropical and subtropical areas of surplus radiation to the high latitude zones of deficit radiation. The transport of heat from equatorial area towards the poles is called '*meridional transport of heat*'.

The meridional transport of heat energy in the form of *sensible heat* is accomplished by the atmospheric circulation and ocean currents which transport heat energy from the 'low latitude surplus energy areas' to the 'high latitude deficit energy areas.' The vertical transport of heat in the atmosphere is accomplished by ascending air in the form of sensible heat and latent heat. Table 4.5 portrays latitudinal transfer of sensible heat.

5

FRONTOGENESIS, CYCLONES AND ANTICYCLONES

FRONTS AND FRONTOGENESIS

Front is that sloping boundary which separates two opposing air masses having contrasting characteristics in terms of air temperature, humidity, density, pressure, and wind direction. An extensive transitional zone between two converging air masses is called frontal zone or frontal surface which represents zone of discontinuity in the properties of opposing contrasting air masses.

Frontal zone is neither parallel nor vertical to ground surface, rather it is inclined at low angle. Though fronts differ from each other in terms of their location, types, and areal extent but they are characterized by the following common characteristics *e.g.* large differences in air temperature across a front, bending isobars, abrupt shift in wind direction, cloudiness and precipitation.

The process associated with the creation of new fronts or the regeneration of decaying fronts already in existence is called frontogenesis. The region having convergence of contrasting air masses is called the region of frontogenesis. The process of destruction or dying of fronts is called frontolysis.

CONDITIONS FOR FRONTOGENESIS

If the distribution of air fronts over the globe is closely observed, it appears that fronts originate only in limited areas which means that fronts originate only when some favourable conditions are available. The necessary conditions for frontogenesis include the presence of two opposing air masses having contrasting properties of air temperature, air pressure, density, humidity and wind direction.

(1) Temperature Difference

Two opposing air masses must have contrasting temperatures *e.g.*, one air mass should be cold, dry and dense while the other should be warm, moist and light. In such conditions, when two air masses converge, then cold and denser air mass invades the area of warm and light air mass and pushes it upward and thus front is formed. It may be pointed out that inspite of convergence of two air masses at the equator no front is formed due to uniform temperature conditions of two air masses (trade winds). A few meteorologists have described the process of frontogenesis near the equator.

(2) Opposite Directions of Air Masses

Convergence of two contrasting air masses is a prerequisite condition for frontogenesis because when two thermally contrasting an masses meet face to face (converge), they try to penetrate into the region of one another and thus a wave-like front is formed. Contrary to this when two air masses di- verge, they move in opposite directions. This situation leads to destruction of fronts, if any. Patterson has identified four types of air circulation wherein only the last two are conducive for frontogenesis.

(i) *Translatory circulation* involves movement of air in horizontal manner from one place to another in the same direction (Fig. 5.1A). This circulation system does not produce temperature variations as the isotherms are parallel and widely spaced. This type of air circulation does not favour the creation of fronts.

(ii) *Rotatory circulation* involves the circulation of air in cyclonic or anticyclonic pattern. In other words, air circulation is circular (Fig. 5.1B and C). Though rotatory air circulation produces temperature variation but front is never created.

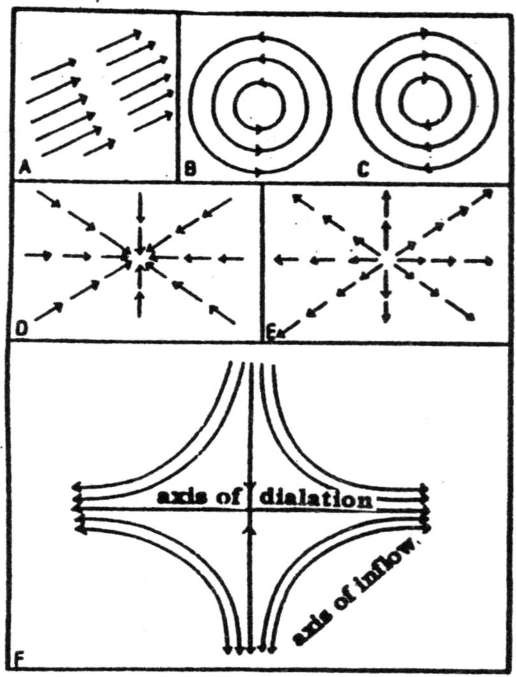

Fig. 5.1 : A = translatory air circulation,
B = rotatory air circulation (cyclonic)
C = rotatory circulation (anticyclonic)
D = convergent circulation,
E = divergent circulation, and
F = deformation circulation.

(iii) *Convergent and divergent circulation :* Convergent circulation involves meeting of winds at common point from all directions (Fig. 5. ID) while divergent circulation refers to spread of winds outward in all directions from a central point (Fig. 5. IE). Though temperature difference is produced in convergent circulation but this situation occurs at a point and not along a line and thus front cannot be created because for frontogenesis it is necessary that temperature difference does exist along a line and not at a point.

(iv) *Deformatory circulation* involves convergence of two contrasting air masses and horizontal spread along a line which is called as axis of outflow or axis of dialation (Fig. 5. IF), while the other axis is called axis of inflow. This type of air circulation is most favourable for frontogenesis.

CREATION OF FRONTS

When two contrasting air masses converge in deformation circulation, they spread horizontally along the axis of outflow or dialation (Fig. 5.2i). In such situation the creation of front depends on the angle between the axis of outflow and isotherms. Fronts do not form when this angle exceeds 45 degrees. As the convergence of air continues, this angle decreases and isotherms try to become parallel to the axis of outflow and frontogenesis is activated. The steepness and intensity of fronts depends on temperature gradient. If two contrasting air masses are parallel to each other and there is no upward displacement of air, stationary front is formed (Fig. 5. 2ii).

Such fronts are climatically insignificant because they are not conducive for cloud formation and precipitation. But such situation is not very common because two contrasting and converging air masses are generally separated by sloping boundary due to deflective force (coriolis force) of the earth and cold and dense air mass pushes warm and light airmass upward (Fig. 5.2iii). It may be pointed out that fronts are not linear between two converging contrasting air masses but are zonal in character having a width of 5 to 80 kilometres.

CLASSIFICATION OF FRONTS

Fronts are classified into four principal types on the basis of their different characteristic features *e.g.*;

(1) warm front,

(2) cold front,

(3) occluded front, and

(4) stationary front.

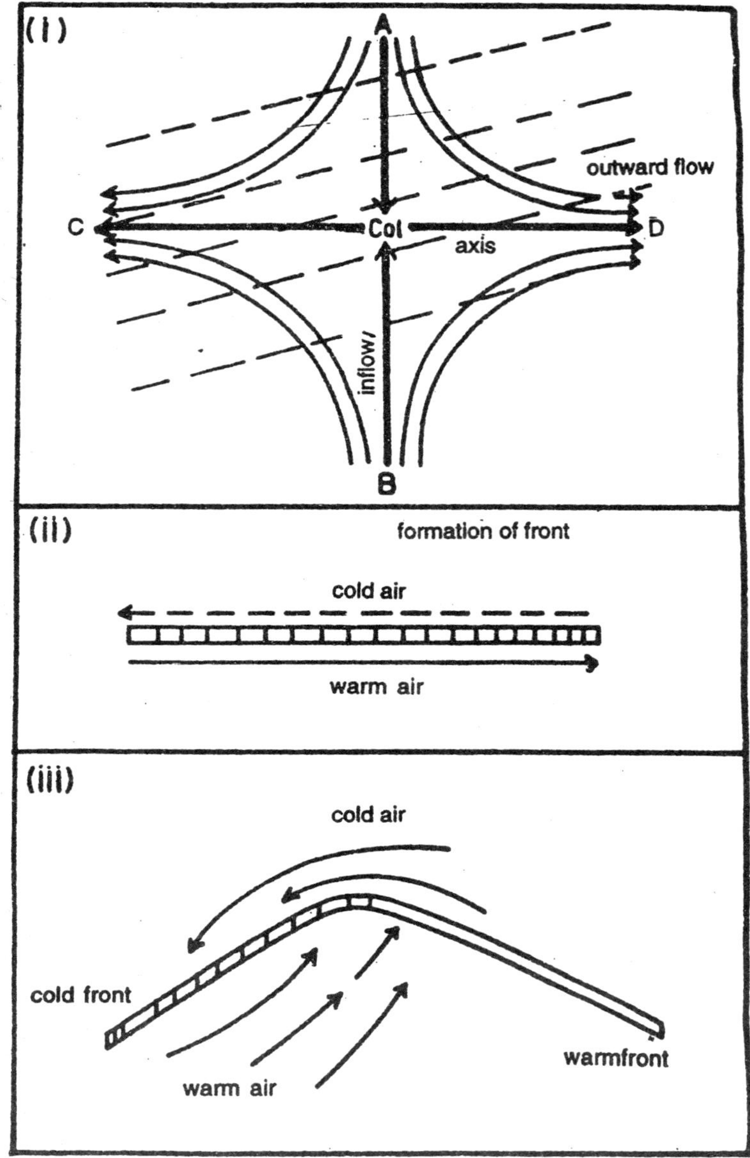

Fig. 5.2 : (i) Frontogenesis, (ii) stationary front, (iii) fully developed front.

(1) Warm Front

Warm front is that gently sloping frontal surface along which warm and light air becomes active and aggressive and rises slowly over cold and dense air (Fig. 5.3A). The average slope of warm fronts in middle latitudes ranges between 1:100 to 1:400.

The gradually rising warm air along the gently sloping warm front is cooled adiabatically, gets saturated and after condensation precipitation occurs over a relatively large area for several hours in the form of moderate to gentle precipitation.

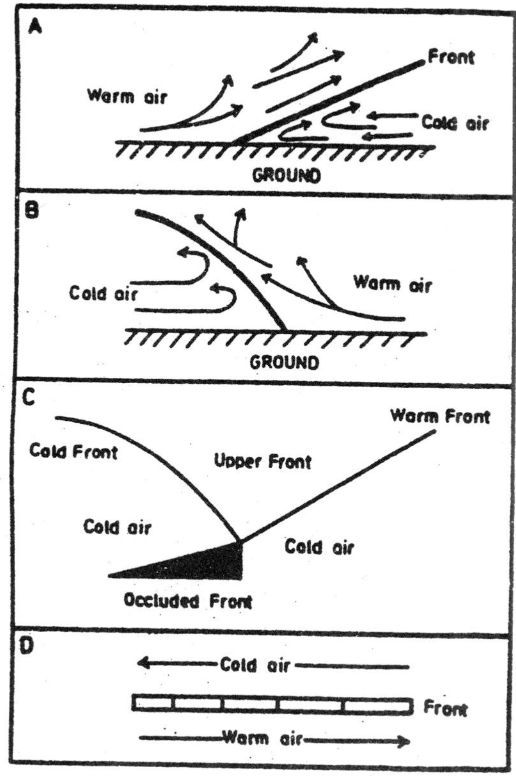

Fig. 5.3 : A = warm front, B = cold front, C = occluded front, and D = stationary front.

(2) Cold Front

Cold front is that sloping frontal surface along which cold air becomes active and aggressive and invades the warm air territory and being denser remains at the ground but forcibly uplifts the warm and light air. Since the air motion is retarded at the ground surface due to friction while the free air above has higher velocity and hence the cold front becomes much steeper than warm front. This is why the slope of cold front varies from 1:50 to 1:100 (which means the rise of the wedge of cold air at the rate of one kilometre for every 50 to 100 kilometres). A cold front is associated with bad weather characterized by thick clouds, heavy downpour with thunder-storms, lightning etc. Sometimes, cold frontal precipitation is also associated with snowfall and hailstorms.

(3) Occluded Front

Occluded front is formed when cold front overtakes warm front and warm air is completely displaced from the ground surface (Fig. 5.3C).

(4) Stationary Front

Stationary front is formed when two contrasting air masses converge in such a way that they become parallel to each other and there is no ascent of air. In fact, the surface position of stationary front does not move either forward or backward.

FRONTAL ZONES

There are two principal zones on the globe where two air masses converge along a line *e.g.* (1) equatorial low pressure belt, and (2) subpolar low pressure belt. It may be noted that fronts are not formed everywhere but they are formed only in those areas where two air masses converge. There are 3 principal frontal zones on the earth's surface *e.g.* (1) polar frontal zone, (2) arctic frontal zone, and (3) intertropical frontal zone.

(1) Polar Frontal Zone

This frontal zone is also known as Atlantic polar front because fronts are formed over the north-western parts of the

Atlantic Ocean. Polar fronts are formed in the middle latitudes in both the hemispheres because of convergence of polar continental air mass (cold and dense air) and tropical maritime air mass (warm and light). These fronts are more developed in the northern parts of the Pacific and Atlantic oceans. They are more active during winter seasons because the temperature contrast is more pronounced between polar continental and tropical maritime air masses. The temperate cyclones produced along polar front travel from west to east under the influence of the westerlies and yield widespread precipitation over large area extending from the eastern parts of North America to the western Europe. Polar front is weakened during summer season because the temperature contrast between two air masses decreases to minimum.

(2) Arctic Frontal Zone

Arctic fronts are created in the arctic areas due to convergence of continental and maritime polar air masses. These fronts are not always very strong and active because of less temperature contrast between two air masses. Active fronts are formed when relatively warm polar maritime air mass meets extremely cold arctic air mass due to which temperature contrasts are accentuated. These fronts develop along the arctic coasts of Eurasia and north America.

(3) Mediterranean Frontal Zone

This frontal zone is formed due to convergence of cold continental European air mass and winter air mass of North Africa along Mediterranean-Caspian Sea line. The winter cyclonic storms developed over Mediterranean Basin move towards east and north-east. Some of these frontal weather disturbances reach Pakistan and north-west India and account for winter precipitation of north India.

(4) Pacific-Arctic Frontal Zone

Fronts develop along a line running from the Rocky mountains to the Great Lakes during winter season. The winter disturbances or cyclones associated with this frontal zone move south-eastward to such locations as Texas and northern Mexico and bring intense cold waves throughout the southern United States.

(5) Pacific Polar Frontal Zone

Two frontal zones develop during winter season over the north Pacific Ocean *e.g.* (i) near northeast Asiatic coast, and (ii) near north-west North American coast. The frontal disturbances developed along the frontal zone of north-east Pacific Ocean (north-west coast of North America) influence the wintertime weather conditions of the western coast of North America extending from the Gulf of Alaska to southern California, and western Mexico. These storms account for winter precipitation of the western coasts of North America.

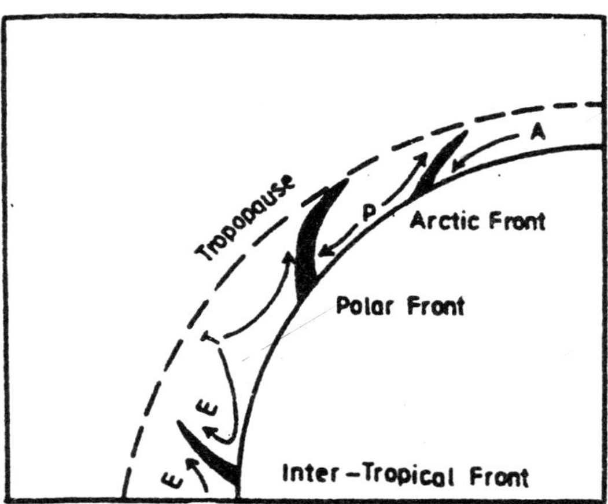

Fig. 5.4 : Major frontal zones over the globe.

(6) Intertropical Frontal Zone

According to some meteorologists tropical front is formed near the equator due to convergence of north-east and south-east trade winds. It may be pointed out that there is no temperature contrast between these two air masses (trades) of similar origin and same properties.

CHANGING WEATHER ASSOCIATED WITH FRONTS

Since fronts are formed due to convergence of two air masses of contrasting temperatures and hence contrasting weather conditions are found from north to south or south to north. Differences in terms of temperature, humidity, precipitation, cloudiness, and wind direction are experienced along different fronts *e.g.* warm and cold fronts.

(1) Weather Associated With Warm Fronts

Warm air becomes active and aggressive along warm front as it invades cold air zone and thus being lighter it gradually rises over cold air and is cooled adiabatically from below. Cooling of warm air causes condensation and cloud formation followed by precipitation. If the aggressive warm air is stable and less humid, condensation occurs at great height and hence much lifting of air is required. On the other hand, if the warm air is moist and unstable, only a slight lifting causes condensation and precipitation.

The warm front precipitation is of long duration, moderate but widespread because of gentle slope of warm front. There are frequent changes in cloud types. The sequence of clouds from above downward comprises cirrus, cirro-stratus, alto-stratum and nimbo-stratus. When warm front advances forward, the warm sector comes over the observation place. There is sudden change in weather conditions with the arrival of warm sector *e.g.* sudden increase in temperature and specific humidity, decrease in air pressure, disappearance of clouds, clear sky, and break in precipitation.

(2) Weather Associated With Cold Front

Cold and dense air becomes active and aggressive along cold front wherein cold air invades warm air region and pushes it upward while it, being denser, settles downward. If cold front passes away soon, weather also becomes clear soon, otherwise if the front becomes stationary, the sky becomes overcast with cumulo-nimbus clouds provided that cold air is moist and unstable, and frontal thunderstorms are formed. Heavy precipitation occurs but is of short duration. The consequent weather is characterized by decrease in air temperature, increase in air pressure, decrease in specific and relative humidity and change in wind direction from 45° to 180°. Precipitation is accompanied by lightning and cloud thunder. Some times, rainfall is associated with hailstorms. After the passage of cold front, clouds disappear, precipitation terminates and weather becomes clear and north-west cold winds set in.

CYCLONES

Cyclones are centres of low pressure surrounded by closed isobars having increasing pressure outward and closed air circulation from outside towards the central low pressure in such a way that air blows inward in anticlockwise in the northern hemisphere and clockwise in the southern hemisphere (Fig. 5.5). Cyclones are also termed as atmospheric disturbances. They range in shape from circular, elliptical to 'V shape. When the velocity of winds increases to such an extent that they attain gale force, the atmospheric disturbance or cyclone is called a cyclonic storm. From the locational point of view cyclones are classified into two principal types *e.g.* (i) extratropical cyclones (also called as temperate cyclones or wave cyclones) and (ii) tropical cyclones.

TEMPERATE CYCLONES

Temperate cyclones, also called as extratropical cyclones or wave cyclones or simply depressions are atmospheric disturbances having low pressure in the centre and increasing pressure outward. They are in fact low pressure centres produced in the middle latitudes characterized by converging and rising air, cloudiness and precipitation.

Because of their varying shapes such as near circular, elliptical or wedge (V) they are variously called as *'low'*, *'depressions'* or *'troughs'*. They are formed in the regions extending between 35°-65° latitudes in both the hemispheres due to convergence of two contrasting air masses *e.g.* warm, moist and light tropical air masses (westerly winds) and cold, and dense polar air masses.

The polar fronts created due to these two opposing air masses are responsible for the origin and development of temperate cyclones. After their formation temperate cyclones move in easterly direction under the influence of westerly winds and control the weather conditions in the middle latitudes.

Types of Temperate Cyclones

Though temperate cyclones are mainly originated due to convergence of two contrasting air masses in terms of temperature, pressure, and humidity but some local cyclones also form due to other reasons related to temperature variations and consequent pressure differences. Based on above considerations temperate cyclones are divided into 3 categories viz. (i) dynamic cyclones, (ii) thermal cyclones, and (iii) secondary cyclones.

(1) *Dynamic cyclones* are, in fact, real temperate cyclones because they are formed due to convergence of cold polar air masses and warm and moist maritime tropical air masses. These cyclones affect the weather conditions of very large areas in middle latitudes. Different fronts (*e.g.* warm front and cold front) and sectors (*e.g.* warm and cold sectors) are fully developed in dynamic cyclones. They are called dynamic because they are dynamically produced *e.g.* due to convergence and invasion of two contrasting air masses into the territories clone another.

(2) *Thermal cyclones*-According to Brunt thermal cyclones are formed due to development of low pressure centres on the continents in summers in temperate regions and as such winds blow from all directions towards the low pressure centres. Such thermally induced temperate cyclones are stationary at their places of origin and different fronts are not developed. Such thermally induced cyclones in the middle latitudes have been named by Humphreys insolation cyclones. According to Humphreys thermal cyclones are produced due to development of low pressure centres over warm water surfaces of seas surrounded by cold land surfaces during winter season. It may be pointed out that both types of cyclones as referred to above are, in fact, thermal cyclones because they are directly related to insolation. The only difference is that they develop over land surfaces in summers (*e.g.* over Iberian Peninsula, Alaska, S. W. USA, and N. W. Australia) and over sea surfaces in winters (*e.g.* over Okhotsk Sea, Norwegian Sea, to the south of Iceland and Greenland etc.).

(3) Secondary cyclones are those which develop due to passage of cold winds over warm sea after the occlusion of main cyclone. They are short-lived and very weak.

Shape, Size and Velocity

Temperate cyclones are of different shapes *e.g.* circular, semi-circular, elliptical, elongated or 'V shaped, but all of them are characterized by low pressure in their centres and closed isobars. The pressure difference between the centre and periphery is about 10 to 20mb but some times it increases to 35mb. It means that pressure increases from the centre towards outer margin. Temperate cyclones also greatly vary in size and extent. Average large diameter of an ideal cyclone is about 1900km (1200 miles) while short diameter measures 1000km (640 miles). It may be pointed out that no two cyclones are identical in terms of their size as their diameters range from 150km to more than 3000km. Some times, temperate cyclones are so large and extensive that they cover an area of 1,000,000 square kilometres. The vertical extent of an average cyclone is about 10-12 km. The temperate cyclones move eastward under the influence of westerly winds with average velocity of 32km per hour in summers and 48km per hour in winters.

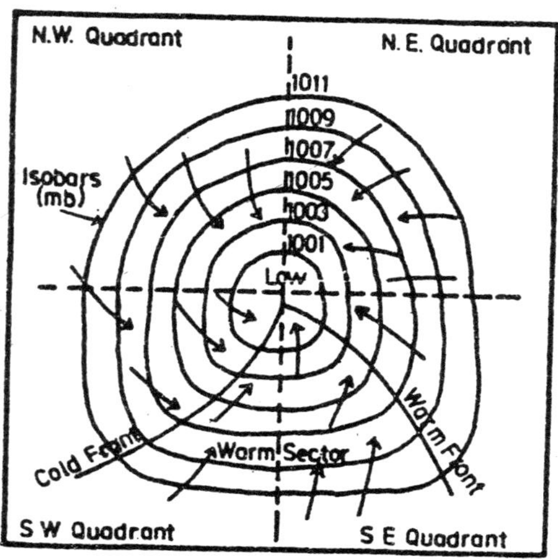

Fig. 5.5 : An average temperate cyclone in northern hemisphere.

Wind System

Since there is low pressure in the centre of temperate cyclone and air pressure increases outward and hence winds blow from the periphery towards the centre but these winds do not reach the centre straight rather they cut the isobars at the angle of 20° to 40° due to friction and coriolis force and thus wind direction becomes anticlockwise in the northern hemisphere and clockwise in the southern hemisphere. The centre-bound inward air circulation becomes of convergent pattern but the winds do not aggregate at the centre but they ascend upward and expand outward so that low pressure centre is always maintained so long as the cyclone is alive. Since temperate cyclones are formed due to convergence of two contrasting air masses (*i.e.*, cold, dry and dense air mass and warm, moist and light air mass) and hence it is natural that there are variations in the nature and direction of winds in different parts of the cyclones. The tropical and subtropical warm and moist air is of generally westerly direction while polar cold air is generally easterly. The convergence of these air masses forms warm front, warm sector, cold front, and cold sector. Before the arrival of warm front the wind direction is easterly but it changes to southerly and southwesterly at the time of the arrival of warm front.

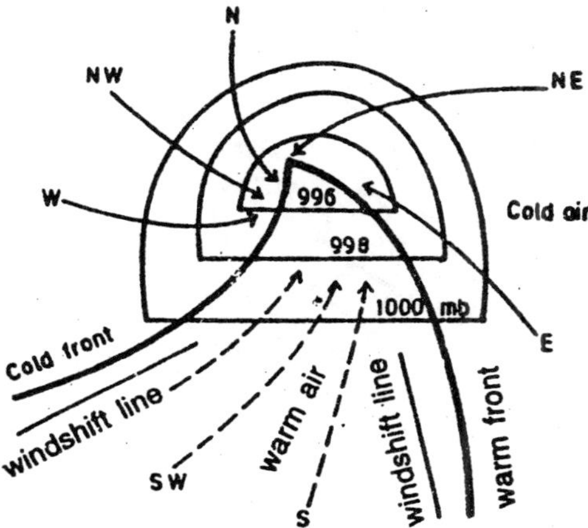

Fig. 5.6 : Wind pattern in temperate cyclone (northern hemisphere).

The warm front and warm sector are, thus, characterized by warm southerly and southwesterly winds while the direction of winds changes to westerly, northwesterly and northerly at the arrival of cold front and cold sector. The cold front and cold sector are characterized by cold winds. It is apparent that there is sudden change in wind direction along the warm and cold fronts. The line along which wind changes its direction is called wind shift line (Fig. 5.6).

Temperature

'Different temperatures are noted in different parts of temperate cyclones because of their origin due to convergence of two thermally contrasting air masses. The southern part of cyclone records higher temperature because of the dominance of warm air while the north-eastern, northern and north-western parts record low temperature because of the dominance of cold polar air mass.

The western part records lowest temperature. The temperature within the cyclones depends on the properties of air masses, general weather conditions and moisture content in the air. Isotherms generally tend in north-northeast to southwest direction (Fig. 5.7) in the northern hemisphere.

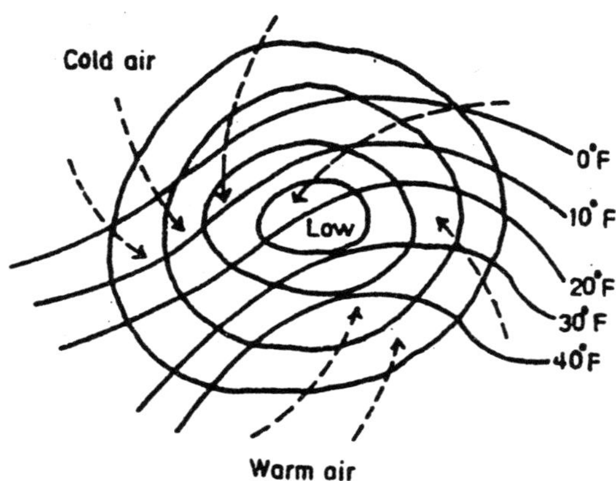

Fig. 5.7 : Temperature and isotherms in temperate cyclones (northern hemisphere).

Source Regions and Tracks of Movement

The areas frequented by temperate cyclones mostly lie in the middle and high latitudes extending between 35°-65° latitudes in both the hemispheres. These cyclones move, on an average, in easterly direction but since their tracks are highly variable and hence cyclonic tracks are always considered in zonal pattern rather than in linear pattern. The paths followed by these cyclones are called *'storm tracks'*. The following are the most favourable breeding areas of temperate cyclones.

(1) Cyclones after originating in the north Pacific off the north-east and eastern coasts of Asia move in easterly and north-easterly direction towards the Gulf of Alaska and ultimately merge with Aleutian Lows from where they follow southerly direction and reach as far south as southern California. The cyclones moving inland dissipate and are occluded at the windward western slopes of the Rocky mountains.

(2) There are four principal areas of frontogenesis in North America *e.g.*

   (a) Area east of Sierra Nevada Range,

   (b) Eastern Colorado, where temperate cyclones are called *Colorado Lows*,

   (c) Area east of Canadian Rocky mountains, where cyclones are known as *Alberta Lows*, and

   (d) Great Lakes region,

(3) The cyclones originating in the Gulf of Mexico follow northerly trajectory to the east of the Appalachians and following the course of the Gulf Stream merge with the Icelandic Zone of frontogenesis.

(4) North-west North Atlantic off the north-east coast of North America- the cyclones originating in this area move in easterly direction and enter the northwestern parts of Europe.

(5) Cyclones originating in the area between Iceland and Barents Sea follow easterly trajectory and affect the weather conditions of north Europe.

(6) There are two main zones of frontogenesis in continental Europe *e.g.* (a) Baltic Sea and (b) Mediterranean Sea. Some of the cyclones originating over the Mediterranean Sea after following easterly direction reach Pakistan and north India in winter season where most of the winter precipitation is received through these storms. Majority of the cyclones of Mediterranean origin move north-east-ward and reach Commonwealth of Independent States (CIS, some Republics of former USSR).

Weather Conditions Associated With Temperate Cyclones

Different parts of temperate cyclones are associated with varying weather conditions because of different types of air masses and varying temperature conditions. The observation point of a moving temperate cyclone experiences different weather conditions at the time of arrival and passage of warm front, warm sector, cold front and cold sector.

(i) Arrival of Cyclone-When the cyclone coming from the western direction draws nearer to the observation point, wind velocity slows down considerably, air pressure decreases and the sun and the moon are encircled by halo which is infact the reflection of thin veneers of cirrus and cirrostratus clouds in the west. Temperature suddenly increases when the cyclone comes very close to the observation point, wind direction changes from easterly to south-easterly, the cloud cover thickens and the sky becomes overcast with dark, thick and low clouds.

(ii) *Warm Frontal Precipitation :* Clouds become very thick and dark with the arrival of warm front of the cyclone and heavy showers begin with nimbostratus clouds. Since the warm air rises slowly along the front, and hence the precipitation is slow, gradual but of long duration. The warm frontal precipitation largely depends on the amount of moisture and instability of the rising warm air. If the air is full of moisture and is unstable, there is sufficient precipitation, the sky is overcast and the sun is not visible for several hours.

(iii) *Warm Sector :* The warm sector comes over the observation point after the passage of warm front and there is sudden change in the pre-existing weather conditions. The wind direction becomes southerly. The sky becomes cloudless and clear. There is sudden rise in air temperature and increase in the specific humidity of the air but air pressure decreases remarkably. Though weather becomes clear but there may be some occasional drizzles. In all, the weather is clear and pleasant.

(iv) *Cold Front :* Temperature registers marked decrease with the arrival of cold front. Cold increases considerably. The cold air pushes the warm air upward and there is change in wind direction from southerly to south-westerly and westerly. Sky is again covered with clouds which soon start precipitation.

(v) *Cold Frontal Precipitation :* Sky becomes overcast with cumulonimbus clouds which yield heavy showers. Since the warm air is forcibly lifted upward hurriedly, the cold frontal precipitation is in the form of heavy downpour with cloud, thunder and lightning but the precipitation is of short duration and less widespread because the cold sector is very close.

(vi) *Cold Sector :* Weather again changes remarkably with the passage of cold front and arrival of cold sector. Sky becomes cloudless and hence clear. There is sharp fall in air temperature and considerable rise in air pressure but decrease in specific humidity. Wind direction changes from 45° to 180° and thus it becomes true westerly. After the occlusion of cyclone the weather conditions of pre-cyclone period again set in.

Origin of Temperate Cyclones

No commonly acceptable theory of the origin of temperate cyclones could be propounded as yet. The first pioneer serious attempt was made by Fitzroy in the year 1863 in this precarious field. He postulated that extratropical or temperate cyclones originated because of the convergence of two opposing air masses of contrasting physical properties (*i.e.*, temperature, pressure, density and humidity). In 1911 Shaw and Lempfert pointed out that temperate cyclones originated due to inflow of winds from all directions towards the centre. The theory of cyclongenesis as propounded by Shaw and Lempfert is known as dynamic theory. At a later date convection current theory was postulated to explain the origin of temperate cyclones but this theory was severely criticised and discarded. Thereafter came eddy theory according to which temperate cyclones originate due to the formation of eddies caused by obstructions in the advancing air masses. Two Norwegian meteorologists, V. Bjerknes and J. Bjerknes put forth 'polar front theory', also knows as wave theory or Bergen theory in the year 1918 to explain the origin of extratropical cyclones. Lately, baroclinic wave theory was formulated for cyclone development on the basis of recent data from middle and upper troposphere derived through satellites and radars.

Polar front theory also called as *'frontal theory'* or *'wave theory'* or Bergen theory as propounded by V. Bjerknes and J. Bjerknes in 1918, is primarily based on the processes of the formation of fronts. It may be pointed out that fronts are formed due to convergence of two air masses of different physical properties coming from opposite directions. One air mass is polar in character and is cold, denser and north-easterly in direction while the other air mass is tropical or subtropical in origin and is warm, moist, lighter and south-westerly in direction (in the northern hemisphere). When these two contrasting air masses converge along a line in the middle latitudes (temperate regions), they move parallel to each other and thus a stationary front is formed (Fig. 5.8A).

No cyclone can develop from such stationary front because there is no vertical movement in the air, rather winds are more or less stable. On the other hand, when two opposing air masses collide against each other and try to attack the territory of one another, unstable waves are formed which help in the origin and development of temper- ate cyclones (Fig. 5.8B). In the beginning the surface separating two air masses (more technically called as surface of discontinuity) is almost straight but it becomes unstable and wave-like when the warm and cold air masses attempt to penetrate in the regions of one another. Such unstable wavy front is called polar front.

When south-westerly warm and moist air mass enters the territory of cold polar air mass along the polar front, it being lighter rises upward, with the result a centre of low pressure is formed. Now winds from all directions rush up towards this centre of low pressure and thus a cyclone is formed. It may be pointed out that the cyclone forming wave developed due to convergence of cold and warm air masses is divided into two parts *e.g.* the eastern part of the wave, where eastward advancing warm tropical or subtropical air mass ascends over a wedge of cold air mass is called warm front (Fig. 5.8C) while the western part, where cold polar air mass pushes warm air mass upward forcibly, is called cold front (Fig. 5.8C).

It is to be remembered that warm air mass is aggressive along the warm front where it overrides cold air mass whereas the cold air mass becomes aggressive along the cold front because it replaces warm air by pushing it upward. The south-western and north-western sectors of the cyclone are called warm sector and cold sector respectively.

The low pressure in the eastern part of the cyclone is intensified with the arrival of warm air. This draws the winds towards the centre from nearby areas, with the result cold front advances more rapidly than warm front. Consequently, cold and warm fronts come close to each other resulting into the destruction of warm front. The cyclone dies due to disappearance of warm front. This process of cyclone destruction is called occlusion (Fig. 5.8E and F). Some times, weak and feable minor cyclone is formed after the occlusion of main cyclone.

Such secondary cyclone is called subcyclone. Secondary cyclone is generally formed when some warm air still remains in the cold front after the occlusion of main front with the result a centre of low pressure is re-established and winds blow towards this centre from all sides. forming a new weak cyclone. It may be pointed out that though the formation and development of temperate cyclones is a quick process but it passes through a series of successive stages. The period of a cyclone from its inception (cyclogenesis) to its termination (frontolysis or occlusion) is called the 'life cycle of cyclone' which is completed through six successive stages (Fig. 5.8).

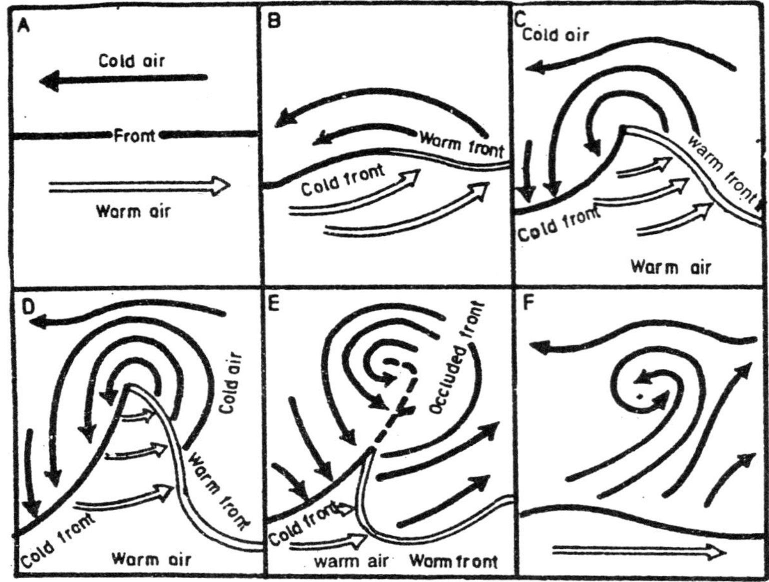

Fig. 5.8 : Stages of life-cycle of a cyclone.

(1) *First stage* involves the convergence of two air masses of contrasting physical properties and directions. Initially, the air masses (warm and cold) move parallel to each other and a stationary front is formed. This is called initial stage.

(2) *Second stage* is also called as 'incipient stage', during which the warm and cold air masses penetrate into the territories of each other and thus a wave-like front is formed.

(3) *Third stage* is the mature stage when the cyclone is fully developed and isobars become almost circular.

(4) *Fourth stage* warm sector is narrowed in extent due to the advancement of cold front than warm front, as cold front comes nearer to warm front.

(5) *Fifth stage* starts with the occlusion of cyclone when the advancing cold front finally overtakes the warm front and an occluded front is formed.

(6) *Sixth stage* warm sector completely disappears, occluded front is eliminated and ultimately cyclone dies out.

TROPICAL CYCLONES

General Characteristics

Cyclones developed in the regions lying between the tropics of Capricorn and Cancer are called tropical cyclones which are not regular and uniform like extratropical or temperate cyclones. There are numerous forms of these cyclones which vary considerably in shape, size, velocity, and weather conditions. The weather conditions of low latitudes mainly rainfall regimes are largely controlled by tropical cyclones. They are characterized by the following salient features—

(1) Size of tropical cyclones varies considerably. On an average, their diameters range between 80km and 300km but some times they become so small that their diameter is restricted to 50km or even less.

(2) They advance with varying velocities. Weak cyclones move at the speed of about 32km per hour while hurricanes attain the velocity of 180km per hour or more.

(3) Tropical cyclones become more vigorous and move with very high velocity over the oceans but become weak and feeble while moving over land areas and ultimately die out after reaching the interior portion of the continents. This is why these cyclones affect only the coastal areas of the continents (*e.g.* south and south-east coasts of the USA, Tamil Nadu, Orissa and West Bengal coasts of India, southern coastal regions of Bangladesh etc.).

(4) The centre of the cyclone is characterized by extremely low pressure. Isobars are more or less circular but are fewer in number. This is why winds hurriedly rush up towards the centre and attain gale velocity.

(5) Like temperate cyclones, tropical cyclones are not characterized by temperature variations in their different parts because they do not have different fronts (warm and cold fronts).

(6) There are no different rainfall cells in the tropical cyclones as is the case of temperate cyclones and hence each part of the cyclones yields rainfall.

(7) Tropical cyclones are not always mobile. Some times, they become stationary over a particular place for several days and yield heavy rainfall causing flood deluge and environmental disaster.

(8) The tracks of tropical cyclones vary considerably in different parts. Normally, they move from east to west under the influence of trade winds. The general direction is westerly upto 15° latitude from the equator, poleward between 15°-30° latitudes, and again westerly. These cyclones weaken when they enter subtropical regions.

(9) Tropical cyclones are confined to a particular period of the year, mainly during summer season. The frequency and affected areas of tropical cyclones are far less than those of the temperate cyclones.

(10) Tropical cyclones become disastrous natural hazards because of their high wind speed of 180 to 400km per hour, high tidal surges, high rainfall intensity (highest recorded rainfall value exceeded 2000mm per day in Philippines), very low atmospheric pressure causing unusually rise in sealevel, and their persistence for several days or say about one week over a particular place.

Types of Tropical Cyclones

It may be pointed out that tropical—cylones are so varied in size, weather conditions and their general characteristics that no two cyclones are identical and therefore it becomes very difficult to classify them into certain categories. Generally, they are divided into 4 major types.

(1) Tropical disturbances or easterly waves

(2) Tropical depressions

(3) Tropical storms

(4) Hurricanes or typhoons

On the basis of intensity they are divided into two principal types and 4 subtypes.

(1) Weak cyclone

(i) Tropical disturbances

(ii) Tropical depressions

(2) Strong and furious cyclones

(iii) Hurricanes or typhoons

(iv) Tornadoes

(1) *Tropical disturbances* are migratory wave-like cyclones and are associated with easterly trade winds. They are also called easterly waves. Winds move towards centre with low speed. Though they move in westerly direction under the influence of trade winds with low velocity but they are most extensive and widespread and influence the weather conditions of both tropical and subtropical areas. Most of the easterly waves develop between 5° and 20° north latitudes in the western parts of the oceans.

Some times, they are so sluggish that they remain stationary over an area for several days. They are associated with heavy cumulus or cumulonimbus clouds which yield moderate to heavy rainfall with thunderstorms. Some times, the easterly waves are so greatly intensified that they develop into hurricanes. Generally, they develop in the Caribbean Sea and North Pacific Ocean during summer months.

(2) *Tropical depressions* are centres of low pressure surrounded by more than one closed isobars and are very small in size. Wind velocity around low pressure centre ranges between 40-50km per hour. Their direction and velocity are highly variable. Some times, they remain stationary at a place for several days. They usually develop in the vicinity of inter-tropical convergence (ITC) but seldom develop in the trade wind belt. Tropical depressions generally influence the weather conditions of India and north Australia during summers. After

being originated in the Bay of Bengal these cyclones move in north-westerly and westerly directions and reach inner parts of India. Some times, they become so strong that they yield heavy downpour resulting into severe floods.

(3) *Tropical storms* are low pressure centres and are surrounded by closed isobars wherein winds move towards the centre with the velocity ranging between 40 to 120km per hour. They frequently develop in the Bay of Bengal and Arabian Sea during summer season. They also develop in the Caribbean Sea and in the vicinity of Philippines. Many of these cyclones become violent and disastrous atmospheric hazards as they cause heavy rainfall and thus inundate lowlying areas of Bangladesh, delta region of West Bengal and coastal areas of Orissa, Andhra Pradhes and Tamil Nadu. The northern parts of Bay of Bengal mostly the Ganga Delta plains of West Bengal, India and Bangladesh very often suffer from frequent severe cyclonic storms and resultant storm surges (tidal waves) because of a combination of several natural conditions and phenomena such as astronomical tides, tunneling coast configuration, low and flat terrains of coastal areas and frequent occurrence of severe cyclonic storms. The most disastrous cyclone, which hit the coastal low land of Bangladesh on November 12,1970, claimed 3.00,000 human lives. Similarly, the deadly cyclone of 1737 claimed the lives of 3,00,000 people in the east coast of India. The disastrous cyclone of 1977 moving with a speed of 175km per hour killed 55,000 people, destroyed the homes of 2,000,000 people and ruined 1,200,000 hectares of agricultural crops and made most of the coastal land barren and wasteland because of deposition of thick layer of salt on the soils by storm surges in Andhra Pradesh. Supercyclone of Orissa of 1999 (Oct. 29-31) with wind velocity of more than 300 km per hour killed about 100,000 people (official figure, 10,000), washed out 200 villages, damaged standing crops of 1.75 million hectares and claimed loss of property worth 1,000 billion rupees in the coastal districts.

(4) *Hurricances or Typhoons* :The extensive tropical cyclones surrounded by several closed isobars are called hurricanes in the USA and typhoons in China. They are also called willy-willy in Australia, cyclones in Indian Ocean, 'baguio' in Philippines, '*taifu*' in Japan etc. Hurricanes are, in fact, most violent, most awesome, and most disastrous hazards of all the atmospheric disturbances. They move with average speed of more than 120km per hour. Though hurricanes are most extensive and violent but their climatic importance is limited because of their fewer numbers and their occurrence in limited areas. Though hurricanes and temperate cyclones look similar in appearance but they may be differentiated on the following grounds.

(1) Hurricanes are represented by more symmetrical and circular isobars. Pressure increases sharply from the centre towards the outer margin resulting into steep pressure gradient. This is why hurricanes move with great force and high speed.

(2) The rainfall occurring from hurricanes is in the form of heavy downpour and is widespread and uniformly distributed whereas precipitation from temperate cyclones is confined to only warm and cold fronts. Warm and cold sectors are devoid of precipitation.

(3) There is no temperature variation in hurricanes. They are also not characterized by different types of fronts (warm and cold fronts) and contrasting air masses as is the case with temperate cyclones.

(4) There is no change in wind direction in hurricanes. Winds blow from the outer margin to- wards the centre and then rise upward.

(5) Hurricanes are not associated with anti-cyclones.

(6) Unlike temperate cyclones they move from east to west.

Besides, hurricanes are characterized by the following properties. The diameters range between 160 and 640km. The size of hurricanes is usually small at their origin points near the equator but the size gradually increases away from the equator. The pressure at the centre ranges between 900 and 950mb which is perhaps the lowest pressure of all the tropical cyclones. The pressure gradient between the centre and outer margin ranges from 10mb to 55mb. The areas of 6 to 48 sq km around the centre of hurricane is generally dry and rainless and winds are feeble. This is called 'eye of the cyclone'. The waves caused in the oceans due to ferocity of hurricanes are called hurricane waves which are generally from 3 to 6m in height. These storm surges inundate the coastal areas with immense volume of oceanic water and thus cause immense loss to human health and wealth. Hurricanes extend upto 12,000m above the ocean surface. They last for many days and some times for more than a week.

Origin of Tropical Cyclones

There is no commonly acceptable viewpoint for the origin of tropical cyclones because the exact mechanism of the formation and development of these cyclones could not be properly understood as yet. According to the advocates of frontal theory all types of cyclones originate because of frontogenesis. Inspite of the absence of two contrasting air masses in the equatorial region fronts are formed due to meeting of land and sea winds. Initially, different fronts are formed but later they disappear. This frontal concept of the origin of tropical cyclone is no longer acceptable because tropical cyclones in no case are related to fronts. In fact, tropical cyclone is like a heat engine which is energised by the latent heat of condensation. On an average, tropical cyclones are formed due to development of low pressure of thermal origin. They develop when the following requirements are fulfilled-

(1) There should be continuous supply of abundant warm and moist air. Without doubt tropical cyclones originate over warm oceans having surface temperature of 27°C during summer season.

(2) Higher value of coriolis force is required for the origin of these cyclones. It is apparent that tropical cyclones are practically absent in a belt of 50-80 wide on both

sides of the equator where coriolis force is minimum. It means that cyclonic circulation of air is caused due to deflection in wind direction resulting from coriolis force. Majority of the tropical cyclones originate within a belt of 5°-200 latitudes in the western parts of the oceans.

(3) They are associated with inter-tropical convergence (ITC) which extends from 5° to 30°N latitudes during summer season.

(4) Pre-existing weak tropical disturbances intensify and ultimately develop into high intensity violent tropical cyclones.

(5) There should be anticylonic circulation at the height of 9000 to 15000m above the surface disturbance. The upper air anticyclonic circulation sucks the air from the ocean surface above and thus the upward movement of air is accelerated and low pressure centre at the surface is further intensified.

(6) Tropical cyclones develop around small atmospheric vortices in the inter-tropical convergence zone (ITC).

Weather Conditions Associated With Tropical Cyclones

The arrival of tropical cyclones at a particular place is heralded by sudden increase in air temperature and wind velocity, marked decrease in air pressure, appearance of cirrus or cirrostratus clouds in the sky, and emergence of high waves in the oceans. The clouds are thickened and become cumulonimbus which yield heavy rains. The clouds are also associated with thunder and lightning. On an average, a single storm yields 100 to 250mm of rainfall but if obstructed by relief barrier it may give as heavy rains as 750 to 1000mm. The visibility becomes zero because the sky is overcast with thick and dark thunder clouds. Such destructive conditions persist for a few hours only. The arrival of the centre or the eye of the cyclone is characterized by calm breezes, clear sky, rainless fine and settled weather, and low pressure at the centre. Such weather conditions do not persist for more than half an hour.

The weather suddenly changes with the arrival of the rear portion of the cyclone as the sky again becomes overcast, wind direction changes, and pressure sharply goes up. There is heavy downpour with cloud thunder and lightning and storm becomes very severe and furious. This situation persists for several hours. Slowly and slowly the ferocity of cyclone starts declining and the weather becomes calm after the cyclone has passed off. The sea surface also becomes calm and clear weather sets in.

Distribution of Tropical Cyclones

Tropical cyclones mostly develop over the ocean surface between 5°-15 latitudes in both the hemispheres and influence weather of coastal areas of the continents. There are 6 major regions of the tropical cyclones *e.g.*;

(1) West Indies. Gulf of Mexico, and Caribbean Sea,

(2) Western North Pacific Ocean including Philippines Islands, China Sea, and Japanese Islands,

(3) Arabian Sea and Bay of Bengal,

(4) Eastern Pacific coastal region off Mexico and Central America,

(5) South Indian Ocean off Madagascar (Malagasi), and

(6) Western South Pacific Ocean, in the region of Samoa and Fiji Island and the east and north coasts of Australia.

North Atlantic Ocean : It may be pointed out that the occurrences of tropical cyclones are rhythemic in nature because they are restricted to a certain season of a year which varies from one region to the other region. On an average, about 7 cyclones develop every year in the southern and south-western parts of the Atlantic Ocean, most of which become hurricanes. They develop;

(i) In August and September around Cape Verde Island,

(ii) Between June and October to the north and east of West Indies and to the south of the Atlantic coast of the USA,

(iii) From May to November in the North Caribbean Sea,

(iv) From June to October in the south Caribbean Sea, and

(v) From June to October in the Gulf of Mexico.

North Pacific Ocean : The cyclones after originating off the western coast of Mexico move north-westward and affect the weather of California. Some times, they also reach Hawaii Island. About 5 to 6 tropical cyclones develop each year between June and November and two of them gain hurricane intensity.

South-West North Pacific Ocean : Normally, tropical cyclones develop in China Sea, off the coasts of Philippines Islands and South Japan between May and December. They have disastrous effects on the eastern coasts of China where they gain the ferocity of typhoons. About 12 typhoons develop every year.

South Pacific Ocean : Tropical cyclones develop to the east of Society Island (east of 180° longitude) during December-April and influence the weather of north-east coast of Australia.

North Indian Ocean : After originating in the Arabian Sea and Bay of Bengal tropical cyclones (also called as depressions) influence the weather conditions of India on a large-scale between April and December.

South Indian Ocean : Cyclones originate off the coasts of Re Union, Madagascar, and Maritius islands between November and April.

Environmental Impact Of Tropical Cyclones

Tropical cyclones are very severe disastrous natural hazards which inflict heavy loss to human lives and property in terms of destruction of buildings, transport systems, water and power supply system, disruption of communication system, destruction

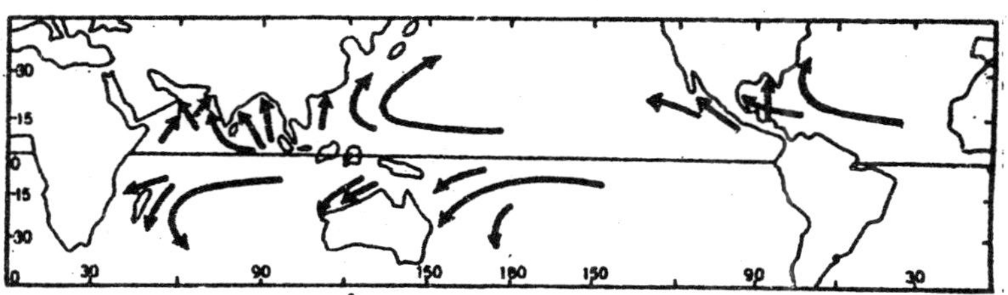

Fig. 5.9 : Tracks of Tropical Cyclones.

of standing agricultural crops, domestic and wild animals, natural vegetation, private and public institutions etc. through damages caused by high velocity winds, floods and storm surges. Tables 5.1 to 5.3 depict the death toll of human life by tropical cyclones in different parts of the world.

Table 5.1 : Some Noteworthy Indian Tropical Cyclones

Year	*Human deaths*	*Year*	*Human deaths*
1731	300,000	1864	50,000
1789	20,000	1977	55,000
1833	50,000	1990	598
1839	20,000	1999	100,000

Note : The intensity of 1990 Andhara cyclone was 25 times greater than the 1977 Andhra cyclone but human causality could be contained because of correct prediction and better warning systems but the property damage could not be stopped.

Table 5.2 : Notable Tropical Cyclonic Disasters in Bangladesh.

Year	*Human deaths*	*Year*	*Human deaths*
1822	40,000	1963	11,468
1876	100,000	1970	300,000
1879	175,000	1976	100,000
1960	5,149	1985	11,000

Table 5.3 : Typhoon Disasters in the Far East.

Year	*Country*	*Human deaths*
1881	China	300,000
1923	Japan	250,000
1950	Japan	5,000

THUNDERSTORMS

General Characteristics

Thunderstorms are local storms characterized by swift upward movement of air and heavy rainfall with cloud thunder and lightning. According to A.N. Strahler "a thunderstorm is an intense local storm associated with large, dense cumulonimbus clouds in which there are very strong updraft of air." Because of heavy downpour associated with thunderstorms they are also called 'cloud bursts' but the rainfall is of very short duration. Thunderstorms differ from cyclones in that the latter are almost circular in shape wherein winds blow from out side towards the centre while the former is characterized by strong updraft of air. They are considered to be special case of convective mechanism.

Structure of Thunderstorms

A thunderstorm consists of several convective cells which are characterized by strong updraft of air. Each cell passes through a phase of life cycle consisting of youth, mature and old stages. It may be pointed out that different cells in a single thunderstorm may be in varying stages. The first stage (youth) is called cumulus stage when warm air rises strongly upward and helps in the formation of clouds. The second stage or mature stage is characterized by both upward and downward movement of winds and occurrence of rainfall. The third stage or senile stage or dissipating stage is characterized by downward movement of winds which spread over the ground surface and stop vertical movement of winds. Clouds spread in the sky in umbrella shape and become altostratus and cirrostratus resulting into dissipation of thunderstorm.

Conditions For Thunderstorm Development

Atmospheric instability, updraft of potentially unstable air, abundant supply of warm and moist air, thick clouds etc. are the factors which favour the development of thunderstorms. The upward movement of warm and moist air is prerequisite condition for the origin of thunderstorms. Surface heating through intense insolation causes convective mechanism resulting into updraft of air and atmospheric instability. This is why thunderstorms originate mainly during summer season, warm days of a season, and warm hours of a day. It appears that warm, moist and rising unstable air is the most important factor in the development of thunderstorms. This becomes possible when normal lapse rate of temperature is greater than adiabatic rate of temperature change. Besides convective mechanism, warm and moist winds also rise and become unstable due to orographic obstacles. The greater the instability of warm and moist air, the greater the intensity and duration of thunderstorms.

There must be greater thickness of clouds between cloud base *i.e.,* level at which condescension and cloud formation begin) and icing level (*i.e.,* the level at which water droplets change into ice particles). The higher the icing level above cloud base, the greater the thickness of clouds and thus the greater the intensity of convection. Since the icing level is at very low height in the middle latitudes, thunderstorms do not develop there. On the other hand, thunderstorms are common features in the weather of low latitudes because of the higher height of icing level and greater thickness of cloud cover.

Thunderstorms And Weather

(i) *Rain fall* in thunderstorms, unlike tropical cyclones, is in the form of heavy downpour with greatest intensity of all other forms of precipitation but is of short duration because of two factors viz. (i) the air rises abruptly with great force due to which there is quick condensation and cloud formation, and (ii) there is abundant absolute humidity due to high rate of evaporation consequent upon very high temperature during summer season. The rainfall of thunderstorm is closely related to its numerous cells. There is maximum rainfall in the centre and minimum at the periphery of each convective cell. Fully developed cell yields rainfall for about an hour whereas weak cell dies out within few minutes.

(ii) *Hailstorms :* When condensation occurs below freezing point, ice particles are formed which range from the size of a pea to large ball. Hail is not associated with every thunderstorm. Not only this, hail is confined to only certain cells of a thunderstorm. Hails fall down on the ground surface when the rising convection currents become weak and feeble. The sudden fall of hails inflicts great damage to human health and wealth, birds and animals and standing agricultural crops.

(iii) *Lightning :* Electrical discharge centres are developed in a mature thunderstorm. The centres of positive and negative electrical charges develop in the upper and lower portions of the clouds respectively with discharge values ranging between 20 to 30 coulombs. Lightning is produced when the electrical potential gradient between the electrical positive and negative charges becomes very steep. According to another view lightning is produced due to splitting of large water drops. Each water drop has positive and negative electrical charges which remain in neutral state when they are evenly balanced. This balance is disturbed due to splitting of these drops resulting into difference in positive arid negative charges.

(iv) Thundering sound is produced due to sudden and rapid expansion of air columns caused by intense heat (10,000°C) resulting from lightning strokes. This deafening noise produced by vibrating pressure wave due to rapid expansion of air column as mentioned above is called cloud thunder.

(v) *Squall :* The downward movement and divergence of cold air at the ground surface is called squall. The velocity of squalls is equal to and some times greater than hurricane velocity and hence they inflict great damage to human structures and vegetation. Squall is produced after the thunderstorm becomes mature and heavy precipitation occurs.

Classification Of Thunderstorms

Thunderstorms are generally classified on the basis of their mode of origin and lifting factors and mechanisms.

(1) *Air mass thunderstorms*

(i) Heat thunderstorms

(ii) Orographic thunderstorms

(iii) Advectional thunderstorms

(2) *Frontal thunderstorms*

(i) Warm front thunderstorms

(ii) Cold front thunderstorms

(1) *Thermal or local thunderstorms* originate due to intense heating of ground surface through insolation and resultant rising thermal convection currents. They are called local because they influence very limited area. In fact, heat thunderstorms are real thunderstorms which originate in the afternoon during summer season and die out by the evening. Heat thunderstorms are more common in the belt of doldrum because large amount of air moisture, high temperature and convergence of winds provide ideal conditions for their origin and development. Heat thunderstorms become stronger and more vigorous if the surface through which they travel is warmer, otherwise they soon are weakened if the surface is less warm. This is why they die out when they pass through water bodies (lakes, rivers, reservoirs etc.) because of no supply of heat from below. Heat thunderstorms also originate in the inner parts of the continents during summer season in middle latitudes. Heat thunderstorms are difficult to be predicted because of their highly uncertain and variable behaviour.

(2) *Orographic thunderstorms* : when warm, moist and unstable air strikes a mountain barrier, it is forced to rise hurriedly along the hillslope. The latent heat of condensation (release of heat after condensation) accelerates the rate of upward movement of the air. This results in the formation of most active and strong thunderstorm which yields copious heavy rains. This is called cloud burst rain. The southwest Indian monsoon winds after striking the hillslopes produce strong thunderstorms which yield more than 12,000mm of annual rainfall at Cherrapunji. Orographic

thunderstorms are more extensive, widespread, and active than heat thunderstorms and thus their forecast is easy.

(3) *Advectional thunderstorms* are produced due to substantial increase in normal lapse rate of temperature and consequent upward movement of unstable air when a cold air underlies a warm air. Such thunderstorms develop during dark nights when the sky is overcast because the upper layers of the clouds are cóoled due to loss of heat through radiation, with the result cool and dense air settles downward and pushes underlying warm and light air upward resulting into convective mechanism in the air.

(4) *Warm frontal thunderstorms* are produced when sea breezes are more humid and unstable. They are not significant because they are very weak storms. Cumulus clouds do nct form because the air ascends slowly.

(5) *Cold front thunderstorms* develop along the cold front of temperate cyclones when cold and dense air pushes warm and moist air upward with great force. Since they are associated with temperate cyclones, they are easily predicted. Cold front thunderstorms may develop at any time of a day or in any season of a year because their origin is not related to the heating of the ground surface.

TORNADO

Tornadoes are funnel shaped storms which are smallest but most violent and disastrous of all the storms. In fact, they are violently rotating column of air having upper portion of funnel shape of cumulonimbus cloud which is attached to the ground by very narrow column of air. The diameter of the funnel increases from 90m in the lower portion to 460m in the upper portion. It is of very dark colour because of dominance of dust, sands, debris and condensed moisture. The centre of tornado is characterized by extremely low pressure, say 100mb less than the outside environment. Because of such steep pressure gradient winds rush up with great force towards the centre having furious velocity of 600 to 800km per hour. Thus, the swiftly inward moving air is caught into a vortex of the storm and is rapidly lifted upward and cools adiabatically and forms thunderstorm. This is why tornadoes are always associated with violent thunderstorms. Tornadoes may 'develop singly or in groups. When tornadoes move in group (called as tornado family), they cause irreparable heavy loss to human health and wealth.

Though exact mechanism of the origin and development of tornadoes is not fully understood but most of the meteorologists believe that they are formed due to violent convection of conditionally and convectively unstable column of ascending air. The origin of tornadoes has also been related to fronts. The upthrusting of warm and moist tropical and subtropical air mass by cold polar air mass along the cold front presents ideal condition for tornado development. Some times, intense local heating of the ground surface causes strong convection which induces ideal condition for the development of tornadoes. According to Californian scientist Vernon J. Rossaw tornadoes develop because of attraction of two cloud masses. Though tornadoes may develop at any time but they are more common during spring and summer season.

Tornadoes are more common in the southern and eastern USA. The approach of tornadoes is heralded by dark and thick clouds in the sky resulting into complete darkness and minimum visibility and low air pressure. The wind blows with hyper velocity which causes cracks in the buildings. The corks of bottles suddenly open up automatically due to sudden change (lowering) of air pressure.

Tornadoes, though smallest in area of all the hazardous atmospheric storms, are very deadly to human lives and property. On an average, the annual toll caused by tornadoes in the USA includes damage to property worth 100,000,000 US dollars and 150 human deaths. The deadliest parts of tornadoes are the tornado missiles (consisting of uprooted trees, their branches, roofs of buildings etc. which are carried away by the dynamic force of winds) which inflict great damage to buildings, other human structures, and human lives. A tornado, for example, at Lubbock (Texas, USA) in 1970 moved a long cylindrical fertilizer tank (3.35m × 12.5m in size with average weight of 11 tons) for a distance of 1.21 km from its original place. It may be pointed out that a tornado becomes disaster only when its funnel of dark clouds moves by touching the ground through narrow column of swiftly moving winds. The tornado outbreak (occurrence of tornadoes in groups involving large numbers) of 60 tornadoes on February 19, 1884 struck the states of Virginia, North Carolina, South Carolina, Georgia, Alabama, Mississippi, Tennessee, and Kentucky between 10A.M. and 12 midnight wherein the most terrific devastation was caused by violent wind storms ever experienced in the USA before this date. Total damages caused by these tornadoes included loss of property worth 3 to 4 million US dollars, death of 800 persons, injuries to 2,500 people, destruction of 10,000 buildings, homeless and destitute people numbering 10,000 to 15,000.

ANTICYCLONES

General Characteristics

Surrounded by circular isobars anticyclone is such a wind system which has highest air pressure at the centre and lowest at the outer margin and winds blow from the centre outward in clockwise direction in the northern hemisphere and anitclockwise in the southern hemisphere (Fig. 5.10). Thus, anticyclones are high pressure systems and more common in the subtropical high pressure belts but are practically absent in the equatorial regions. They are generally associated with rainless fair weather. This is why anticyclones are called weatherless phenomena. They are characterized by the following properties—

(1) They are usually circular in shape but some times they also assume 'V' shape. There is maximum air pressure at the centre and it decreases outward. The difference of pressure between the centre and periphery of anticyclone ranges between 10-20mb and some times it becomes 35mb.

(2) They are much larger in size and area than temperate cyclones as their diameter is 75 per cent larger than that of the latter.

(3) Though anticyclones follow cyclones but their track is highly variable and unpredictable. They move very sluggishly and some times they become stationary over a particular place for few days. The average velocity of anticyclones is 30-50km per hour.

(4) Because of high pressure at the centre winds blow outward clockwise in the northern hemisphere and anticlockwise in the southern hemisphere.

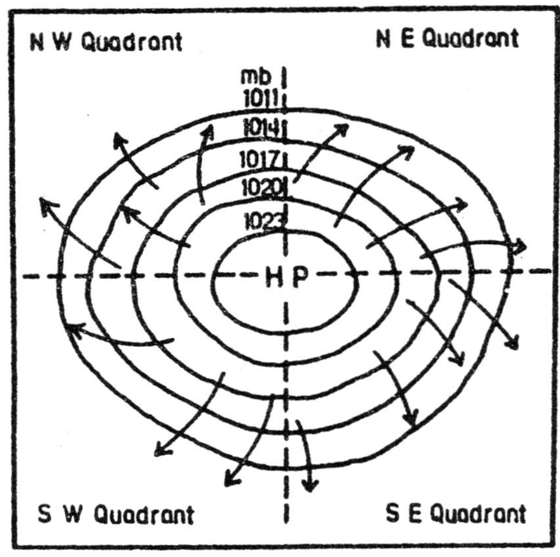

Fig. 5.10 : Air pressure and wind system in an anticyclone.

(5) Winds descend from above at the centre and thus weather becomes clear and rainless because the descending winds cause atmospheric stability.

(6) Temperature in anticyclones depends on weather, nature of air mass and humidity in the air. They record high temperature during summer season due to development of warm air masses whereas they carry low temperature during winter season due to polar cold air masses.

(7) Anticylones do no have fronts.

Wind System

Wind system is not fully developed in anticyclones because of weak pressure gradient. On an average, wind circulation is of divergent system wherein winds spread in all directions from high pressure centre to low pressure periphery. There is westerly wind in the front portion of an eastward advancing cold anticyclone while the rear portion is characterized by easterly winds (Fig. 5.10). The winds are very much sluggish in the rear portion in comparison to the winds in the front portion. The centre is characterized by light breeze. Wind system is seldom developed fully in warm anticyclones.

Shape And Size

Anticyclones are generally of circular, near circular or wedge shape but are very large in size. Some times, they become so large in size that their diameters become 9,000km. There is little difference between the length and width of anticyclones. Temperate anticyclones are so extensive that a single anticyclone covers nearly half of the USA.

Temperature

Anticyclones are originated due to the descent of either polar cold air mass or warm tropical air mass. It is, thus, obvious that cold anticyclones are associated with extremely low temperature and they cause cold waves during winter season but when they come in summer season, weather becomes pleasant. On the other hand, warm anticyclones bring heat waves during summer season in the subtropical regions.

Weather Conditions

Generally, anticyclones are rainless and sky is free of clouds because of the fact that descending air in the centre of anticyclone is warmed up at dry adiabatic rate due to subsidence. This causes rise in temperature which reduces normal lapse rate of temperature, with the result the stability of air increases resulting into marked increase in the aridity of air. This is why anticyclones are indicative of dry weather. This does not mean that anticyclones are always rainless. While passing over oceans some times they pick up moisture and yield light rains with moderate clouds. The arrival of anticyclones is heralded by clearing of clouds, if already present in the sky, clear weather and decrease in wind velocity. The weather of Canada, USA, and north Eurasia is mostly affected by anticyclones.

Types

Anticyclones were classified into (i) warm anticyclones, and (ii) cold anticyclones by Hanzilk in 1909, while Humphreys divided them into 3 types viz. (i) mechanical or dynamic anticyclones, (ii) thermal or heat anticyclones, and (iii) radiation anticyclones. A.E.M. Gaddes has classified anticyclones into 3 types *e.g.* (i) large anticyclone, covering the whole of the continent, (ii) temporary anticyclone, which has a diameter of 250 to 300km and covers only the marginal areas of the continents, and (iii) cyclone-originated anticyclone, which develops due to high pressure caused between two temperate cyclones. Normally, anticyclones are divided into (i) cold anticyclones, (ii) warm anticyclones, and (iii) block to anticyclones.

(1) *Cold anticyclones :* after originating in the arctic regions cold anticyclones advance in easterly and south-easterly directions. Though they are smaller than warm anticyclones in size but move more rapidly than the latter. They are of very low thickness. Very few cold anticyclones are higher than 3,000m. Harwitz and Noble have observed upper atmospheric low pressure above surface anticyclones at US-Canada border but such situation is not a permanent feature. Cold anticyclones are divided into two subtypes *e.g.* (i) temporary anticyclones, which die out in the transit while moving forward, only a few reach tropical regions, and (ii) semipermanent anticyclones, which cover longer distances and are more active. Cold anticyclones are thermally induced because they do not develop due to descent of air from above. They are originated due to development of high pressure because of very low insolation during winter season in the arctic regions. Cold anticyclones follow two tracks, (i) Anticyclones

after originating in the north of Canada move in easterly and south-easterly direction and affect the weather conditions of Canada and USA. (ii) Anticyclones originating in the north of Siberia move towards China, Japan and Alaska. Anticyclones affecting north-west Europe originate with temperate cyclones (in their rear portion). While entering tropical region cold anticyclones die out due to increase in temperature.

(2) *Warm anticyclones* originate in the belt of subtropical high pressure where winds diverge in opposite directions. Thus, warm anticyclones are originated due to descent of air from above and consequent divergence at the surface. They, thus, are dynamically induced. They are large in size but are very sluggish in movement. Some times, they become stationary over a place for several days and weeks. They are associated with light wind, cloudless sky and clear weather. Warm anticyclones mostly influence the weather of S.E. USA and western Europe.

(3) *Blocking anticyclones* develop due to obstruction in the air circulation in the upper troposphere. This is why they are called blocking anticyclones. They develop over N.W. Europe and adjoining Atlantic Ocean and the western part of the N. Pacific Ocean between 140°-170°W longitudes. They are similar to warm anticyclones as regards wind system, air pressure and weather but are small in size and move very slowly.

6

TEMPERATURE

INTRODUCTION

Here, temperature means atmospheric temperature. Indirectly the sun is the major source of atmospheric temperature. In fact, the atmosphere receives very low amount of heat energy from the sun as it receives most of its energy from the longwave terrestrial radiation. Atmosphere receives energy from the sun and the earth in different ways. On an average, the heating and cooling of the atmosphere is accomplished through direct solar radiation and through transfer of energy from the earth through the processes of conduction, convection and radiation. It may be pointed out that 'heat' and 'temperature' generally appear synonym to common people but these two terms differ considerably as regards their real meaning. Heat is a form of energy while temperature denotes the intensity of hotness or coldness of any substance. In other words, heat denotes the quantity of energy present in any substance while temperature refers to the degree of hotness of that subject. Thus, the temperature of any substance may be raised or lowered with the addition or substraction of heat. The transfer of heat from one body to the other is accomplished through conduction, convection, and radiation.

HEATING AND COOLING OF THE ATMOSPHERE

The solar energy received by the earth's surface including both ground surface and water surface (of the seas and the oceans) is convened into heat energy in the form of sensible heat (heat that can be measured by thermometer) and is temporarily stored. This stored energy is radiated from the ground and water surface in the form of longwaves into the atmosphere. The process of radiation of heat energy from the earth's surface is called ground radiation (including radiation from both, ground surface and water surface). The part of this ground radiation after being absorbed by the atmosphere is again radiated back to the earth's surface. This process of radiation of terrestrial heat energy from the atmosphere back to the earth's surface is called counter-radiation. The counter-radiation is effected mainly by water vapour and atmospheric carbon dioxide.

1. Heating of the Atmosphere by Direct Insolation

The heat energy is radiated from the outer surface of the sun (photosphere) in the form of longwaves. The atmosphere absorbs 14 per cent of incoming short wave solar radiation through ozone, water vapour etc. present therein. Seven per cent of this energy is spread in the lower atmosphere up to the height of 2km. It is apparent that this amount is too low to heat the atmosphere significantly.

2. Conduction

The transfer of heat through the molecules of matter in any body is called conduction. The transfer of heat under the process of conduction may be accomplished in two ways viz.:

(i) from one part of a body to the other part of the same body, and

(ii) from one body to the other touching body.

Conduction may be effective only when there is difference in temperatures in different parts of a single body or in two bodies and the process continues till the temperatures of all parts of a body or of two touching bodies become same. It is obvious that heat moves from warmer body to the cooler body through molecular movement. The rate of transfer of heat through molecular movement depends on the heat conductivity of the substance. The substance or a body which allows transfer of heat through conduction at a very fast rate is called good conductor of heat while the substance or a body which retards conduction of heat is called bad or poor conductor of heat. Metal is a good conductor of heat while air is very poor conductor of heat. The earth's surface is heated during day-time after receiving solar radiation. The air coming in contact with the warmer ground surface is also heated because of transfer of heat (conduction of heat) from the ground surface through the molecules to the air. Since air is very poor conductor of heat and hence the transfer of heat from the ground surface through conduction is effective only upto a few metres in the lower atmosphere. The ground surface becomes colder than the air above during winter nights and thus heat is transferred from the lower portion of the atmosphere to the ground surface and thus the atmosphere is cooled.

3. Terrestrial Radiation

The process of transfer of heat from one body to the other body without the aid of a material medium (*e.g.* solid, liquid or gas) is called radiation. There are two basic laws which govern the nature of flow of heat energy through radiation.

(a) *Wien's displacement law* 'states that the wavelength of the radiation is inversely proportional to the absolute temperature of the emitting body'.

(b) *Stefan-Boltzmann law* 'states that flow, or flux of radiation is proportional to the fourth power of the absolute temperature of the radiating body'.

The earth's surface after receiving insolation from the sun through shortwave electromagnetic radiation gets heated and radiates heat to the atmosphere in the form of longwave or infrared radiation throughout 24 hours. It may be remembered that the atmosphere is more or less transparent for incoming shortwave solar radiation but it absorbs more than 90 per cent of outgoing longwave terrestrial radiation through water vapour, carbon dioxide, ozone etc. Thus, the terrestrial radiation is the most important source of heating of the atmosphere. The process of radiation of heat from the earth's surface is called *ground radiation*. The part of this ground radiation after being absorbed by the atmosphere is radiated back to the earth's surface. This process of radiation of terrestrial heat energy from the atmosphere back to me earth's surface is called *counter-radiation* which is effected mainly by water vapour and atmospheric carbon dioxide. This mechanism known as greenhouse effect keeps the lower atmosphere and the ground surface relatively warmer. Thus, the atmosphere acts as window glasspane which allows the shortwave solar radiation to come in and prevent the longwave terrestrial radiation to escape into space.

It is obvious that the increase in the concentration of carbon dioxide in the atmosphere will increase the greenhouse effect and thus the temperature of the earth's surface would increase. It may be pointed out that carbon dioxide also absorbs longwave terrestrial radiation and helps in keeping the lower atmosphere and the ground surface warmer. Water vapour absorbs both the incoming shortwave solar radiation and outgoing longwave terrestrial radiation. Since most of water vapour is concentrated in the lower atmosphere (90 per cent of the total atmospheric water vapour is found upto the height of 5km in the lower atmosphere) and hence both the incoming solar radiation and outgoing terrestrial radiation increase with increasing height. This is the reason that high mountains are called radiation windows.

4. Convection

The transfer of heat energy through the movement of a mass of substance from one place to another place is called convection. The process of convection becomes effective only in fluids or gases because their internal mass motion activates convection of heat energy. The earth's surface gets heated after receiving heat energy (insolation) from the sun. Consequently, the air coming in contact with the warmer earth's surface also gets heated and expands in volume. Thus, warmer air becomes lighter and rises upward and a vertical circulation of air is set in.

Conversely, the relatively colder air aloft becomes heavier because of contraction in volume and thus descends to reach the earth's surface. The descending air is wanned because of dry adiabatic rate and warm ground surface. This warm air again ascends because of increase in volume and decrease in density. The whole mechanism of ascent of warmer air and descent of colder air generates convection currents in the lower atmosphere. This convective mechanism transports heat from the ground surface to the atmosphere and thus helps in the heating of the lower atmosphere. Similarly, horizontal convection currents are also generated on the ground surface.

Besides, atmosphere is also heated through latent heat of condensation, and expansion and compression of air.

MEAN TEMPERATURES

All types of average temperatures *e.g.* monthly, seasonal and annual means, depend on daily temperatures and their averages. The earth receives energy (insolation) through the electromagnetic shortwaves radiated from the photosphere (outer surface of the sun) of the sun during day time. This radiant energy is transformed into heat. The earth's surface receives maximum energy at 12 noon but the maximum temperature never occurs at 12 noon because the transformation of solar energy into heat requires time. The energy received by the earth from solar radiation continues to exceed the energy lost by outgoing longwave radiation from the earth's surface from 6 A.M. to 2-4 P.M. Thus, the curve of air temperature also rises upto 4 P.M. This is why maximum temperature is recorded between 2 P.M.- 4 P.M. The highest temperature recorded within 24 hours is called maximum daily temperature. On the other hand the loss of energy through longwave radiation from the earth's surface exceeds the energy received from the sun from 4 P.M. to the sunrise and hence there is gradual fall of the curve of temperature but the lowest (minimum) temperature is recorded between 4-5 A.M. instead of midnight. The lowest temperature within 24 hours is called *minimum daily temperature*. Thus, there is no coincidence between the time of maximum and minimum amount of insolation received from the sun and maximum and minimum temperatures of the air. This is called *lag of temperature*. The average of maximum and minimum temperatures within 24 hours is called *mean daily temperature*. Similarly, the mean of daily maximum and minimum temperatures of a month is called mean monthly maximum temperature and mean monthly minimum temperature. Mean of maximum and minimum temperatures of 12 months is called mean annual maximum and mean annual minimum temperature. The difference between the maximum and minimum temperatures of a day (24 hours) is called daily or diurnal range of temperature. Similarly, the differences between the monthly maximum and minimum temperatures on one hand and between annual maximum and minimum on the other hand are called monthly range and annual range of temperatures respectively.

DISTRIBUTION OF TEMPERATURE

The spatial and temporal distribution of temperatures is very significant because different types of weather, climates, vegetation zones, animals and human life etc. basically depend on the distribution of temperature, whether horizontal or vertical. The study of distribution of temperature is attempted in several ways viz.:

(i) Temporal distribution, and

(ii) Spatial distribution.

The spatial distribution is studied in two ways *e.g.*:

(a) vertical distribution, and

(b) horizontal distribution of temperature.

The horizontal distribution of temperature is also termed as regional distribution of temperature. The-spatial distribution of temperature is controlled by a variety of factors *e.g.* latitudes or distance from the equator, height from sea level or altitude, distance from the sea coast, nature of land and water, properties of ground surface, nature of slopes, prevailing winds, oceans currents etc.

Factors Controlling the Distribution of Temperature

As elaborated above the following factors control the distribution of temperature on the earth's surface :

(1) Latitudes

The temperature of the atmosphere of a particular place near the ground surface depends on the amount of insolation received at that place. Since the amount of insolation received by the ground surface decreases poleward from the equator *i.e.*, from low latitudes towards high latitudes because the sun's rays become more and more oblique (slanting) poleward and hence air temperature also decreases poleward. It may be noted that though sun's rays are almost vertical over the equator throughout the year but there is no maximum temperature on it rather maximum temperature is recorded along 20°N latitude in July because major portion of insolation is reflected by clouds and sizeable amount of heat is lost in evaporation in the low latitude zone (equatorial zone).

(2) Altitude

The temperature decreases with increasing height from the earth's surface at an average rate of 6.5°C per kilometre because of the following reasons.

(i) The major source of atmospheric heat is the earth's surface from where heat is transferred to the atmosphere through the processes of conduction, radiation and convection. Thus, the portion of the atmosphere coming in direct contact with the earth's surface gets more heat from the ground surface than the portion lying above because as we ascend higher in the atmosphere the amount of heat to be transported above decreases and hence temperature decreases aloft,

(ii) The layers of air are denser near the earth's surface and become lighter with increasing altitudes. The lower layer of air contains more water vapour and dust particles than the layers above and hence it absorbs larger amount of heat radiated from the earth's surface than the upper air layers.

(3) Distance From the Coast

The marine environment moderates the weather conditions of the coastal areas because there is more mixing of temperatures of the coastal seas and oceans and coastal lands due to daily rhythm of land and sea breezes. Thus, the daily range of temperature near the coastal environment is minimum but it increases as the distance from the sea coast increases. Minimum daily range of temperature is the characteristic feature of marine climate while extremely high daily range of temperature characterizes continental climate.

(4) Nature of Land and Water

The contrasting nature of land and water surfaces inflation to the incoming shortwave solar radiation largely affects the spatial and temporal distribution of temperature. It may be pointed out that land becomes warm and cold more quickly than the water body. This is why even after receiving equal amount of insolation the temperature of land becomes more than the temperature of the water body. The following reasons explain the differential rate of heating and cooling of land and water.

(i) The sun's rays penetrate to a depth of only 3 feet in land because it is opaque but they penetrate to greater depth of several metres in water because it is transparent to solar radiation. The thin layer of soils and rocks of land, thus, gets heated quickly because of greater concentration of insolation in much smaller mass of material of ground surface. Similarly, the thin ground layer emits heat quickly and becomes colder. On the other hand, the same amount of insolation falling on water surface has to heat larger volume of water because of the penetration of solar rays to greater depth and thus the temperature of ground surface becomes higher than that of the water surface though the amount of insolation received by both the surfaces may be equal.

(ii) The heat is concentrated at the place where insolation is received and there is very slow process of redistribution of heat by conduction because land surface is static. It may be noted that downward distribution of temperature in the land surface within a day (24 hours) is effective upto the depth of only 10 centimetres. Thus, the land surface becomes warm during day and cold during night very rapidly. On the other hand, water is mobile. The upper surface of water becomes lighter when heated by insolation and thus moves away horizontally to other places and the solar rays have to heat fresh layer of cold water. Secondly, heat is redistributed in water bodies by sea waves, ocean currents and tidal waves. All these extend the period of warming of water surface.

(iii) There is more evaporation from the seas and the oceans and hence more heat is spent in this process with the result oceans get less insolation than the land surface. On the other hand, there is less evaporation from the land surface because of very limited amount of water.

(iv) The *specific heat* (the amount of heat needed to raise the temperature of one gram of a substance by 1°C) of water is much greater than the land because the relative density of water is much lower than that of land surface. It means more heat is required to raise the temperature of one gram of water by 1°C than one gram of land. More specifically, the heat required to raise the temperature of one cubic foot of water by 1°C is two times greater than the heat required for the equal volume of land (one cubic foot). It is apparent that same amount of insolation received by same mass of water and land would increase the temperature of land more than the temperature of equal mass of water.

(v) The reflection (albedo) of incoming solar radiation from the oceanic water surface is far more than from the land surface and thus water receives less insolation than land.

(vi) Oceanic areas are generally clouded and hence they receive less insolation than land surface. But clouds absorb outgoing terrestrial radiation and counter-radiate heat back to the earth's surface. This process retards the loss of heat from the oceanic surfaces and hence slows down the mechanism of cooling of the air lying over the oceans. On the other hand, land surfaces receive more insolation at faster rate because of less cloudiness and simultaneously lose more heat through outgoing terrestrial radiation very quickly.

(5) Nature of Ground Surface

The nature of ground surface in terms of colour, vegetation, land use practices etc. affects the distribution of temperature.

The snow covered surfaces receive very low amount of insolation because they reflect 70 to 90 per cent of incoming shortwave solar radiation and thus polar and arctic areas are characterized by extremely low temperature through- out the year. On the other hand, sandy surfaces record high temperature during day time in the tropical and subtropical areas because they absorb most of solar radiation very quickly as they reflect only 20 to 30 per cent of solar radiation. The following is the albedo of different kinds of surfaces.

Table 6.1 : Albedo of different kinds of surfaces

Surface	*Per cent of radiation being reflected*	*Surface*	*Per cent of radiation being reflected*
Snow cover	70-90	Forest	5-10
Sand	20-30	Water	3-5
Grass	14-37	Thick cloud	70-80
Dry ground	15-25	Thin cloud	25-50
Wet ground	10	Black soils	8-14

Dark-coloured ground surfaces receive more solar radiation than light- coloured surfaces. The temperature of ground surface of forested areas never becomes very high because sizeable amount of heat is spent in the process of evapotranspiration from plant leaves.

(6) Nature of Ground Slope

The ground slope facing the sun receives more insolation because the sun's rays reach the surface more or less straight and hence sun-facing ground surfaces record higher temperature than the leeward slopes where sun's rays reach more obliquely. In the northern hemisphere the southward facing slopes of east-west stretching mountains receive greater amount of insolation than the north-ward facing slopes because of their exposure to the sun for longer duration. This is why most of the valleys situated on the southern slopes of the Alpine mountains have settlements and cultivation.

(7) Prevailing Winds

The prevailing winds help in redistribution of temperature and in carrying moderating effects of the oceans to the adjacent coastal land areas. The winds blowing from low latitudes to high latitudes raise the temperature of the regions where they reach while winds blowing from high latitudes to low latitudes lower the temperature. The winds blowing from oceans to coastal lands bring in marine effects and thus lower the daily range of temperature. The winds coming from higher parts of the mountains lower the temperature in the valleys. The temperature rises at the time of arrival of temperate cyclones while it falls sharply after their passage or their occlusion. The winds associated with warm oceanic currents raise the temperature of coastal lands while the winds coming in contact with cold currents lower the temperature of affected coastal lands. For example, the North Atlantic Drift (extension of Gulf Stream) raises the temperature of coastal areas of north-western Europe and Labrador cold current lowers the temperature in the neighbourhood of New foundland. Some times, local winds change the temperature dramatically. For example, the warm *chioook winds* raise the temperature by 30 to 40°F within a few minutes along the eastern slopes of the Rockies in the USA and the snow melts as by magic at the arrival of chinook. This is why pastures are open throughout the year along the eastern slopes of the Rockies. *Loo*, a local hot wind, raises the temperature in the Ganga Plain of north India to such an extent (maximum temperature ranges between 40°C to 48°C during the day) that heat waves prolong for several days in continuation and several people die of sun stroke.

(8) Ocean Currents

The warm ocean currents flowing from tropical areas to temperate and cold zones raise the average temperature in the affected areas. For example, Gulf Stream raises the average temperature of the coastal areas of north-western Europe while Kuroshio warm current raises the temperature of Japanese coasts. Labrador, Peru, California, Kurile etc. cold currents lower the temperatures of affected areas. In fact, ocean currents moving from one place to another equalize the temperatures.

VERTICAL DISTRIBUTION OF TEMPERATURE

Temperature decreases with increasing height in the troposphere but the rate of decrease varies according to seasons, duration of sunshine and location. On an average, the rate of decrease of temperature with increasing altitudes in a stationary column of air with absence of any vertical motion is 6.5°C per 1000 metres. This decrease of temperature is called vertical temperature gradient or normal *lapse rate* which is 1000 times greater than the horizontal lapse rate (decrease of temperature with increasing latitudes). The decrease of temperature upward in the atmosphere proves the fact that the atmosphere gets heat from the earth's surface through the processes of conduction, radiation and convection. It is, thus, obvious that as the distance from the earth's surface (the source of direct heat energy to the atmosphere) increases (*i.e.*, as the altitude increases), the air temperature decreases. The following are the reasons for decrease of temperature with increasing altitudes in the troposphere.

(i) Heat is transferred to the atmosphere from the earth's surface through the processes of conduction, radiation and convection. Thus, as the altitude increases the amount of heat transported upward decreases. Consequently, every air layer receives less heat than the air layer lying below.

(ii) The air pressure is higher in the lower portion of the atmosphere near the earth's surface because of weight of all the air layers lying above and thus the air density is maximum in the lower atmosphere but it decreases rapidly upward and the air becomes thin.

(iii) The quantity of water vapour, dust particles, water droplets, carbon dioxide etc., which absorb outgoing longwave terrestrial radiation, is more concentrated in the lower portion of the atmosphere and decreases rapidly with increasing altitude. Thus, the temperature of lower atmosphere becomes more than the air layers lying above because of more and more absorption of terrestrial radiation in the lower air layers. In other words, the temperature decreases upward because of decrease of absorption of terrestrial radiation with increasing height in the troposphere.

It may be pointed out that the decrease of temperature with increasing height is confined to the troposphere only. The height of troposphere is 16km and 6 km over the equator and the poles respectively but this height also varies in different seasons *i.e.*,. it becomes higher in summer than in winter. The upper limit of troposphere is called tropopause.

It is interesting to note that the temperature at tropopause increases from over the equator towards the poles because the height of tropopause decreases from over the equator towards the poles. The height of tropopause during July and January over the equator is –17km while temperature at the top of tropopause is –70°C. The height of tropopause decreases to 15km in July and 12.5km in January over 45°N latitude but the temperature at the top of tropopause increases to –60°C in July and –58°C in January. The height of tropopause further decreases to 10km in July and 9 km in January over the poles but the temperature increases to –45°C in July and –58°C in January.

Upward from tropopause the temperature is reported to increase with increasing height in the stratosphere wherein it becomes 0°C or 32°F at the height of 50km from sea level. This is the upper limit of the stratosphere and is called stratopause. Temperature again decreases with increasing height in the mesosphere (50km–80km). The temperature becomes –80°C at mesopause, the upper limit of the mesosphere.

Beyond mesopause temperature again increases with increasing height in the thermosphere. It is estimated that the temperature at its upper limit (height undecided) becomes 1700°C. It may be pointed out that this temperature cannot be measured by ordinary thermometer because the gases become very light due to very low air density.

Sometimes, temperature increases with increasing height in the troposphere. In other words, some time warm air lies over cold air. This phenomenon is called *inversion of temperature*.

INVERSION OF TEMPERATURE

(1) Meaning of Temperature Inversion

As stated above temperature decreases with increasing altitudes in the troposphere at an average rate of 6.5°C per 1000 metres (normal lapse rate) but sometimes this normal trend of decrease of temperature with increasing heights is reversed under special circumstances *i.e.*, temperature increases upward upto a few kilometres from the earth's surface. This is called *negative lapse rate*. Thus, warm air layer lies over cold air layer. This phenomenon meteorologically is called inversion of temperature. Such situation may occur near the earth's surface or at greater height in the troposphere but the inversion of temperature near the earth's surface is of very short duration because the radiation of heat from the earth's surface during daytime warms up the cold air layer which soon disappears and temperature inversion also disappears. On the other hand, upper air temperature inversion lasts for longer duration because the warming of cold air layer aloft through terrestrial radiation takes relatively longer period of hours.

(2) Types of Temperature Inversion

Inversion of temperature occurs in several conditions. Some times, it occurs at ground surface while some times, it occurs at greater height. Sometimes, it is caused in the static atmospheric conditions while some times, it occurs due to horizontal or vertical movement of air. Thus, temperature inversion is classified into the following types on the basis of relative heights from the earth's surface at which it occurs and the type of air circulation.

(1) Non-Advectional Inversion

(i) Ground or surface inversion or radiation inversion

(ii) Upper air inversion

(2) Advectional Inversion

(i) Frontal inversion or cyclonic inversion

(ii) Valley inversion due to vertical air movement.

(iii) Surface inversion due to horizontal air movement

(3) Mechanical Inversion

(i) Subsidence inversion

(ii) Turbulence and convective inversion

(1) *Ground surface inversion* also called as *radiation inversion* occurs near the earth's surface due to radiation mechanism. This is also called as non-advectional inversion because it occurs in static atmospheric condition characterized by no movement of air whether horizontal or vertical (it may be noted that air is never static). Such inversion normally occurs during the long cold winter nights in the snow-covered regions of the middle and high latitudes. In fact, surface inversion is caused due to excessive nocturnal cooling of the ground surface due to rapid rate of loss of heat from the ground through outgoing longwave terrestrial radiation. Thus, the air coming in contact with the cool ground surface also becomes cold while the air layer lying above is relatively warm. Consequently, temperature inversion develops because of cold air layer below and warm air layer above (Fig. 6.1).

The ground surface inversion occurs under the following conditions :

(i) Long winter nights, so that the loss of heat by terrestrial radiation from the ground surface during night may exceed the amount of insolation received from the sun through incoming shortwave electromagnetic radiation waves and thus the ground surface becomes too cold.

(ii) Cloudless and clear sky, so that the loss of heat through terrestrial radiation proceeds more rapidly without any obstruction. Clouds absorb terrestrial radiation and hence retard loss of heat from the earth's surface.

(iii) Presence of dry air near the ground surface, so that it may not absorb much heat radiated from the earth's surface as moist air is capable of absorbing much of the radiant heat from the earth's surface.

(iv) Slow movement of air, so that there is no transfer and mixing of heat in the lower layers of the atmosphere.

(v) Snow-covered ground surface, so that there is maximum reflection of incoming solar radiation. Snow, being bad conductor of heat retards the flow of heat from the ground surface lying below the snow-layers to the lower atmosphere.

Since such inversion of temperature occurs in the static atmospheric condition (very little movement of air) and hence it is also called *static* or *non-advectional in version*. Ground inversion occurs in the low latitude areas (tropical and subtropical areas) during winter nights only and the inversion generally disappears with sunrise but some times it persists upto noon. The inversion occurs upto the height of 30-40 feet in the low latitudes, a few hundred feet in the middle latitudes and half a mile in the high latitudes. It is apparent that the duration and height of surface inversion increase poleward. This inversion promotes stability in the lower portion of the atmosphere and causes dense fogs.

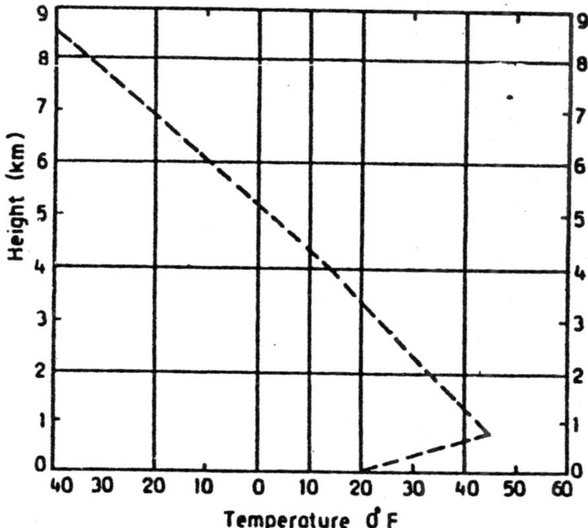

Fig. 6.1: Ground surface inversion of temperature.

(2) *Upper air inversion* is of two types viz. (i) thermal upper air inversion and (ii) mechanical upper air inversion. The thermal upper air inversion is caused by the presence of ozone layer lying between the height of 15 to 35km (even upto 80km) in the stratosphere. The ozone layer absorbs most of the ultraviolet rays radiated from the sun and thus the temperature of this layer becomes much higher than the air layers lying above and below ozone layer. This inversion occurs only when there is no vertical movement of air (either ascent or descent of air). The mechanical inversion of temperature is caused at higher heights in the atmosphere due to subsidence of air and turbulence and convective mechanism. Such inversion occurs in a number of ways *e.g.*

(a) Some times, warm air is suddenly transported upward (due to eddies formed by frictional forces) to the zone of cold air and thus cold air being denser lies under the warm air and inversion of temperature is caused,

(b) The descending air is wanned at the dry adiabatic rate of 10°C per 1000m because of compression. This situation causes the existence of warm air above the cold air. Such mechanical inversion caused by the subsidence of air currents is generally associated with the anticyclonic conditions. Such inversion of temperature is of very common occurrence in the middle latitudes where high pressures are characterized by sinking air. The poleward regions of trade winds are also characterized by high pressure caused by the subsidence of air and thus there is also frequent mechanical inversion of temperature. Since the temperature inversion causes stability in the atmosphere and thus there is dry condition. This is the reason that the poleward parts of trade winds are characterized by arid conditions.

(3) *Advectional inversion of temperature* is also called as dynamic inversion because it is always caused due to

either horizontal or vertical movements of air. Strong wind movement and unstable conditions of the atmosphere are prerequisite conditions for advectional inversion of temperature. This type is further divided into 3 subtypes on the basis of the nature of air movements *e.g.* : (i) frontal or cyclonic inversion, (ii) valley inversion due to vertical movement of air, and (iii) surface inversion due to horizontal movement of air.

(i) *Frontal or cyclonic inversion* is caused in the temperate zones due to temperate cyclones which are formed due to the convergence of warm westerlies and cold polar winds in the northern hemisphere. The warm air is pushed up by the cold polar air and thus the warm air overlies the cold air because it is lighter than the cold air. Thus, the existence of warm air above and cold air below reverses the normal lapse rate (decrease of temperature with increasing height) and inversion of temperature occurs. It may be pointed out that contrary to other types of temperature inversion, the inversion layer associated with frontal or cyclonic inversion is always sloping because the coriolis force causes the boundary zone (front) between the warm westerlies and cold polar air masses to become slopy. It is also interesting to note that air moisture increases upward in frontal inversion of temperature while it decreases upward in other types of temperature inversion.

(ii) *Surface inversion* of temperature caused by horizontal movement of air occurs in several situations. Such inversion is caused when warm air invades the area of cold air or cold air moves into the area of warm air because warm air being lighter is pushed upward by relatively denser cold air. When the warm air moves, such inversion is caused over the continents during winter and over the oceans during summer but when the cold air becomes active and invades the areas of warm air, such inversion occurs over the continents during summer and over the oceans during winter. Such surface inversion occurs generally in the low latitudes. The convergence of cold and warm ocean currents also causes such inversion of temperature.

(iii) *Valley inversion* generally occurs in the mountainous valleys due to radiation and vertical movement of air. This is also called vertical advectional inversion of temperature. The temperature of the upper parts of the valleys in mountainous areas becomes exceedingly low during winter nights because of rapid rate of loss of heat from the surface through terrestrial radiation. Consequently, the air coming in contact with the cool surface also becomes cool. On the other hand, the temperature of the valley floor does not fall considerably because of comparatively low rate of loss of heat through terrestrial radiation. Thus, the air remains warmer than the air aloft and hence the warm and light air of the valley floor is pushed upward by the descending cold and heavier air of the upper part of the valley.

Thus, there is warm air aloft and cold air in the valley floor and inversion of temperature is caused. This situation is responsible for severe frost in the valley floors causing great damage to fruit orchards and vegetables and agricultural crops whereas the upper parts of the valleys are free from frost. This is why the valley floors are avoided for human settlements while the upper parts are inhabited in the mountainous valleys of middle latitudes. Fig. 6.2 explains the mechanism of valley inversion of temperature.

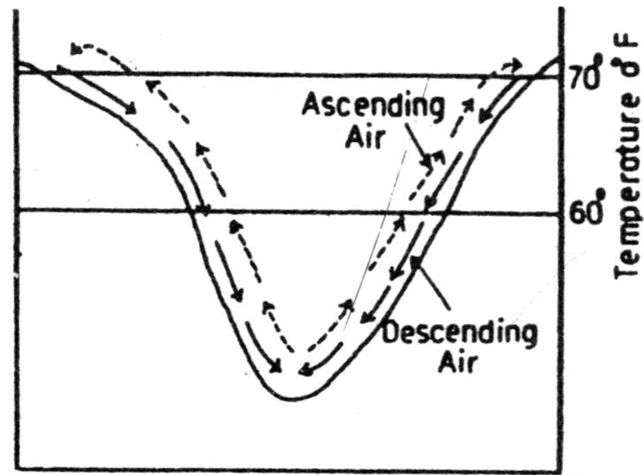

Fig. 6.2 : Valley inversion of temperature.

Significance of Inversion of Temperature

Though, inversion of temperature denotes local and temporary conditions of the atmosphere but there are several climatic effects of inversion which are of great significance to man and his economic activities.

(1) Fog is formed due to the situation of warm air above and cold air below because the warm air is cooled from below and resultant condensation causes the formation of tiny droplets around suspended dust particles and smokes during winter nights. The smokes coming out of houses and chimneys intensify fogs and become responsible for the occurrence of urban smogs. When smog is mixed with air pollutants such as sulphur dioxide it becomes poisonous and deadly health hazard to human beings. The incidents of deadly urban smogs of December 1930 in Meuse Valley of Belgium (600 people fell ill and 63 people died), of October 26, 1948 at Donora in Pennsylvania (USA), and of 1952 in London (causing 4000 human deaths) tell the ordeal of urban smogs. Fogs are also formed at the convergence of warm and cold ocean currents. For example, dense and extensive fogs are formed near Newfoundland due to convergence of warm Gulf Stream and cold Labrador current. Fogs reduce atmospheric visibility and thus they are responsible for several cases of accidents of air crafts while taking off and landing and ships in the oceans. Road and rail transport is also badly affected by the occurrence of dense fogs. Though, generally fogs are unfavourable for many agricultural crops such as grams, peas, mustard plants, wheat etc. but some times they are also favourable for some crops such as coffee plants In Yemen hills of Arabia where fogs protect coffee plants from direct strong sun's rays.

(2) Inversion of temperature causes frost when the condensation of warm air due to its cooling by cold air below occurs at temperature below freezing point. Frost is definitely economically unfavourable weather phenomenon mainly for crops because fruit orchards and several agricultural crops such as potatoes, tomatoes,

peas etc. are totally damaged over-night. There are frequent cases of destruction of fruit orchards in the lower parts of valleys due to severe frost caused by inversion of temperature in California (USA). Fruit orchards were severely damaged in California in the year 1913, 1917 and 1923 due to severe frosts. The valley floors in the hills of Brazil are avoided for coffee cultivation because of frequent frosts. Alternatively, coffee is planted on the upper slopes of the valleys. The upper parts of the valleys are inhabited in Switzerland while lower parts are avoided. In fact, the upper slopes of valleys are characterized by fruit orchards, hotels and motels (for tourists), various resort centres etc. while the lower slopes are deserted.

(3) Inversion of temperature causes atmospheric stability which stops upward (ascent) and downward (descent) movements of air. The atmospheric stability discourages rainfall and favours dry condition. The inversion of temperature caused by the subsidence of air resulting into anticyclonic conditions increases aridity. This is why the western parts of the continents situated between 20°-30° latitudes and characterized by anticyclonic condition represent most widespread tropical deserts of the world.

HORIZONTAL DISTRIBUTION OF TEMPERATURE

On an average, temperature decreases from the equator towards the poles and thus low latitudes are characterized by highest temperature whereas high latitudes record lowest temperature. It may be noted that highest temperature is never recorded at the equator instead it is recorded near the tropics of Capricorn and Cancer because a sizeable portion of incoming solar radiation is reflected by clouds and a large part of heat is spent in the process of evaporation near the equator. On the other hand, most of the incoming solar radiation reaches the ground surface due to clear sky near the tropics of Cancer and Capricorn. Normally, temperature decreases from equator towards the poles. This change of temperature rather decrease of temperature poleward is called *temperature gradient*. The rate of decrease of temperature between the tropics of Cancer and Capricorn is rather low and therefore temperature gradient is insignificant but temperature decreases more rapidly from the tropics poleward and therefore temperature gradient is also very high. The horizontal distribution of temperature is represented and studied with the help of isotherms.

Isotherms

The lines drawn on maps joining the places of equal temperature reduced to sea level are called *isotherms.* It is necessary to reduce the actual temperatures of all places at sea level before drawing isotherms. It is, thus, obvious that isotherms do not represent the real temperature of the places through which they pass rather they show temperature of the places at sea level. This is why the isotherm maps are not useful for farmers because they need real temperature of a particular face for growing crops. Normally, isotherms run east-west and are generally parallel to latitudes.

This trend shows strong control of latitudes on the horizontal distribution of temperature. Generally, isotherms are straight but they bend at the junction of continents and oceans due to differential heating and cooling of land and water. Isothermal lines are more irregular in the northern hemisphere because of large extent of continents but they are more regular in the southern hemisphere due to over-dominance of oceans. Isotherms are generally closely spaced in the northern hemisphere but they are widely spaced in the southern hemisphere. The closely spaced isotherms denote rapid rate of change of temperature and steep temperature gradient. On the other hand, widely spaced isotherms indicate slow rate of temperature change and low temperature gradient. On an average, isotherms trending from land towards the ocean bend equatorward during summer and poleward during winter. On the other hand, isotherms trending from the oceans to the continents bend poleward during summer and equatorward during winter. The isotherms during the months of January and July are taken as representatives for the study of horizontal distribution of temperature during winter and summer seasons respectively because they represent seasonal extremes.

July and January

If we look at the world map representing horizontal distribution of temperature in January and July (Figs. 6.3 and 6.4) the following facts come to light:

(1) The months of July and January are wannest and coldest in the northern hemisphere whereas the warmest and coldest months in the southern hemisphere are January and July respectively.

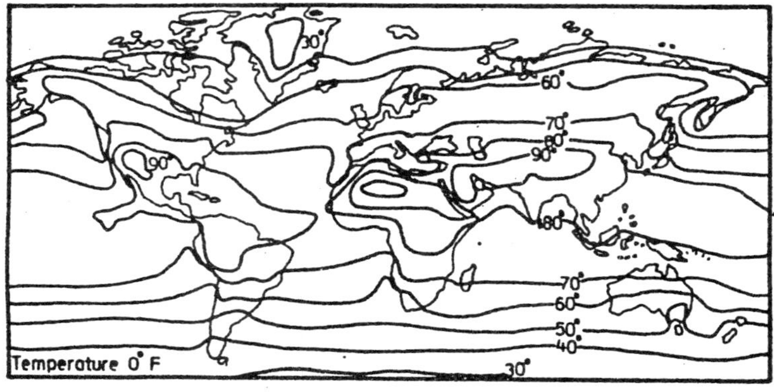

Fig. 6.3 : Isotherms representing horizontal distribution of temperature in July.

(2) Both the maps (Figs. 6.3 and 6.4) show latitudinal shifts of isotherms in accordance with seasonal shifting of overhead sun but this shifting of isotherms is more pronounced on the continents.

(3) The maximum temperatures in January and July are always recorded on the continents. Minimum temperature in January is observed in Asia and North America.

(4) January isotherms suddenly bend poleward while passing through warm portions of the oceans and bend equatorward while passing through the cold portions of the oceans in January in the northern hemisphere while the trend is opposite in July, On the other hand, the isotherms are more or less regular and straight in the

southern hemisphere because of over-dominance of oceans.

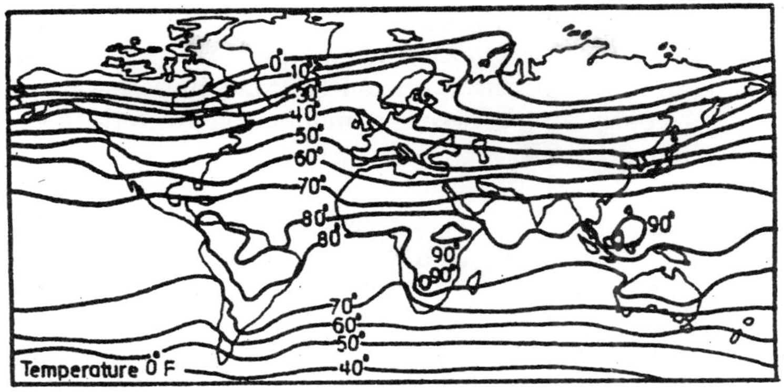

Fig. 34.4 : Isotherms representing horizontal distribution of temperature in January.

(5) Temperature gradient is more pronounced during winter than summer.

REGIONAL DISTRIBUTION OF TEMPERATURE

According to ancient Greek thinkers the globe is divided into three temperature zones on the basis of latitudes (Fig. 34.5) *e.g.*:

(1) Tropical zone,

(2) Temperate zone, and

(3) Frigid zone.

(1) *Tropical zone* extends between the tropics of Cancer (23.5°N) and Capricorn (23.5°S). The Sun's rays are more or less vertical on the equator throughout the year. The remaining areas are also characterized by vertical sun's rays at least once every year.

There is no winter around the equator because of high temperature prevailing throughout the year but as one approaches the tropics of Cancer and Capricorn summer and winter are clearly observed and differentiated.

(2) *Temperate zone* extends between 23.5° and 66.5° latitudes in both the hemispheres. Though, the duration of day and night is longer in this zone but it is never more than 24 hours. There are marked seasonal contrasts with the northward and southward (summer and winter solstices) migration of the overhead sun and thus the range of temperature between summers and winters becomes exceptionally very high.

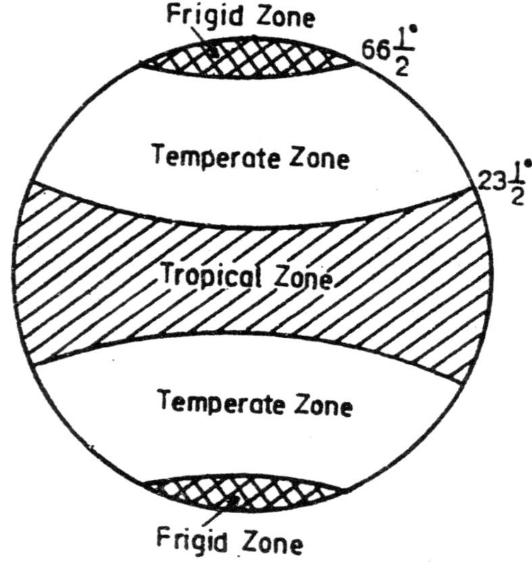

Fig. 6.5 : Temperature zones according to the views of ancient Greek thinkers.

(3) *Frigid zone* extending between 66.5° latitude and the poles in both the hemispheres is characterized by more oblique sun's rays throughout the year resulting into exceptionally very low temperature. The length of day and night is more than 24 hours. Days and nights are of 6 months duration at the poles. Sun is never vertical and the ground is covered with snow as temperature more or less remains below freezing point.

It may be pointed out that the Greeks gave undue importance to latitudes in determining different temperature zones and overlooked the controls of contrasting nature of continents and oceans in terms of their heating and cooling, prevailing winds, ocean currents, nature of ground surface etc. Taking all these factors in consideration Soupan divided the globe into temperature zones on the basis of isotherms. According to him the outer limit of tropical zone should be determined on the basis of annual isotherm of 68°F (20°C). The boundary between temperate and frigid zones in the northern hemisphere should be demarcated by 50°F (10°C) isotherm of July while January isotherm of 50°F(10°C) should separate temperate zone from the frigid zone in the southern hemisphere.

WORLD TEMPERATURE DISTRIBUTION

The seasonal distribution of temperature forms one criterion in the delimitation of climatic types. Sea-level isothermal maps show for the northern hemisphere in winter a marked northward bend in the isotherms over the sea and a southward bend over the land; trace particularly the course of the 0°C (32°F) isotherm. In summer the isotherms bend northward (but not so markedly) over the land. In the southern hemisphere, where the land areas are small, the parallel east-west trend of the isotherms is more in evidence, particularly south of 35°S.

The highest shade temperature recorded in the world was 58°C (136-4°F) at Azizia in Tripoli (1922) and again in San Luis Potosi in Mexico (1933), the highest in Britain 38°G (100-5°F) at Tonbridge, Kent (1868). Conversely, the world's lowest recorded shade temperature was –88°C (–127°F) at Vostok in the Antarctic (1960), the lowest in Britain –27.2°C (–17°F) at Braemar (1895).

'*Sensible temperatures*' are the sensations of heat and cold felt by the human body. Temperature in this connection is intimately related to humidity; dry heat mitigates high temperatures by accelerating evaporation from the skin surface, while moist heat makes the body feel distinctly uncomfortable. An index of this is the wet-bulb *thermometer*; if the wet-bulb reads much above 24°C (75°F), sustained manual labour is extremely difficult. Damp raw cold is much more trying than dry cold; damp air is a better conductor of heat than dry air, and allows the escape of heat from the body in cold weather.

Very low temperatures can be endured (as in central Canada) when the air is dry and still and the sun shining. A *temperature-humidity index* can be constructed by various empirical formulae which affords an indication of the effects on human comfort; on one such scale the index of 60 to 65 represents ideal conditions, while 80 can be regarded as so uncomfortable (as happens in New York and elsewhere) offices and factories may be closed. Another physiological index in cold climates is *windchill*, obtained from a formula involving both temperature and wind force.

These facts are of prime importance to the geographer; they involve problems of tropical and low-altitude settlement, of white acclimatization in the tropics, of the exploitation of the Northland, and of the efficiency of labour.

Accumulated Temperatures

The calculation of accumulated temperatures is a statistical device for indicating the accumulation of warmth over a basic critical value for a certain period of time. This has considerable significance in the study of relationships between plant activity and temperature conditions during the period of growth, in the classification of climatic types, and in the analysis of climatic cycles. The unit used is called either the degree-day or day-degree, which gives an exact measure of the departure of the mean daily figures from the selected base-temperature. Thus, if 6°C is the datum, and 9°C is the mean for a specific month, it will count as $3 \times 31 = 93$ degree-days towards the final total. More accurately, though more tediously, the accumulated temperature for a specific day may be given in degree-hours.

SUNSHINE

The duration of sunshine is partly a function of latitude, for the hours of light (that is, of possible sunshine) vary with the seasons in different latitudes. It is also a function of daytime cloudiness.

Measurement and Recording

The duration of bright sun-shine is measured by means of a Campbell-Stokes recorder, a solid sphere of glass 10 cm in diameter. This focuses the rays of sunshine on to a sensitized card graduated in hours, and so burns a line during the time the sun is shining. Faint sunlight, near dawn or dusk, or when the sun is partially obscured, is not recorded. When mean figures have been obtained over the requisite number of years, the values for each station can be plotted, and lines of equal mean duration of sunshine (isohels) can be interpolated.

Distribution of Sunshine

The sunniest parts of the earth are the hot deserts in the Trade Wind belt. At Helwan Observatory, in the Nile valley south of Cairo, the mean sunshine for the year is 3668 hours, which represents about 82 per cent of the possible amount for that latitude. From June to September the duration is about 90 per cent of the possible amount, and even the cloudiest month (January) receives 70 per cent of the possible. The lands around the Mediterranean Sea and other parts of the world with broadly similar climates enjoy an extremely sunny summer, with a mean of about 90 per cent of the possible duration. Even in winter, the rainy season, days without any sun-shine are few, since the rain tends to be shortlived.

SCALE OF TEMPERATURE

The temperature can be measured in terms of the expansion or contraction in length of a column of liquid as the result of heating or cooling. These changes have been standardized in the form of certain scales; on the *Fahrenheit* scale the boiling-point of water is fixed at 212°, its freezing-point at 32°. The equivalent points on the *Centigrade* (since 1948 officially called Celsius) scale are 100° and 0°, on the Reaumur scale 80° and 0°. Scientists sometimes use the *Absolute* or *Kelvin* scale, in which temperature is measured in degrees Centigrade from the point at which molecular movement ceases, –273·15°C = 0°K.

Instruments

The column of liquid in a thermometer, the instrument used for measuring temperature, is of either mercury or alcohol. Thermometers must be carefully sited; very different readings will be obtained on the ground ('grass temperatures'), at various depths beneath the surface (usually 10 cm, 20 cm, 30 cm and 1 m), on a brick wall, or in the shade. The records usually required are of actual air temperatures—that is, 'shade temperatures'—measured under standardized conditions with precautions to exclude the direct rays of the sun; these are the figures generally quoted by climatologists. The thermometer is placed in a meteorological screen (a wooden box with a double top and bottom and louvred sides, in which the bulb of the thermometer is 1.25 m (4 ft) from the ground). The screen is carefully sited away from buildings.

Maximum and *minimum thermometers* are of great value, in which small dumb-bell-shaped markers indicate the highest and the lowest temperatures attained since the last setting, which is done each 24 hours with a small magnet. Most convenient is the thermograph, the principle of which is the expansion and contraction of a bi-metallic strip in the shape of a coil; one end is fixed, the other actuates a pen which traces a continuous record on a chart fixed to a revolving drum. Another type of thermograph is the Bourdon tube, used in aircraft, in which a curved metal tube is filled with spirit; its curvature changes with the temperature, and these changes are recorded on a chart or dial.

Presentation of Temperature Records

Temperature figures may be presented in several ways. Extreme data are sometimes interesting (as when the Meteorological Office announces that it has been the warmest day in January for seventy years). Extreme maxima and minima, the incidence of frost, and other crucial figures are also of value.

The temperature figures of most value to climatologists seeking to present a long-term view of average conditions are means, with varying corrections and weightings. Diurnal, monthly and annual means are all used, but the most useful are the mean temperatures of the warmest and coolest months respectively over a minimum thirty years of observation. From these can be computed the mean range, that is, the difference between the means of the warmest and coolest months. Annual means are of little value, as an examination of the following mean monthly figures for Peking will show:

J	*F*	*M*	*A*	*M*	*J*	*J*	*A*	*S*	*O*	*N*	*D*
–5	–a	5	14	20	24	26	24	20	12	3	–3°G
23	29	41	57	88	76	79	76	68	54	38	27°F

The range is 31°C (56°F), indicative of the great seasonal climatic contrasts in northern China. The annual mean is 12° C (53°F), yet ten of the twelve months have means far from this figure. The annual mean for Verkhoyansk in Siberia is –16° C (2-7°F), but this is of little significance when we learn that the means of the coldest and warmest months are –51°C (–58°F) and 16°C (60°F).

Temperature Graphs and Maps

Monthly mean temperature for a station can be represented in the form of a simple graph (Fig. 6.6).

The distribution of mean air temperature over a considerable area is shown by drawing an isotherm map. All available stations are located on a base-map, temperatures plotted, and then lines drawn to represent selected temperatures. Rarely will they pass through a station; usually they must be interpolated proportionally. The values plotted may be either actual means, or reduced to sea-level by adding a correction for the altitude of the station; thus a map 'either of 'actual isotherms' or 'sea-level isotherms' can be constructed. The disadvantage of the first is that it closely resembles a contour map, of the second that it portrays a hypothetical state of affairs.

Another very revealing temperature map is one which depicts the difference between the mean temperature (reduced to sea-level) of a station and the mean temperature for all stations in that latitude. This difference will give either a positive anomaly (*i.e.*, it is warmer than the average for the latitude) or a negative anomaly. If these anomalies are plotted, and lines of equal anomaly (isanomals) are interpolated, the map will show clearly such features as the winter cold of continental interiors, the summer heating of the land-masses, and the effects of the oceans.

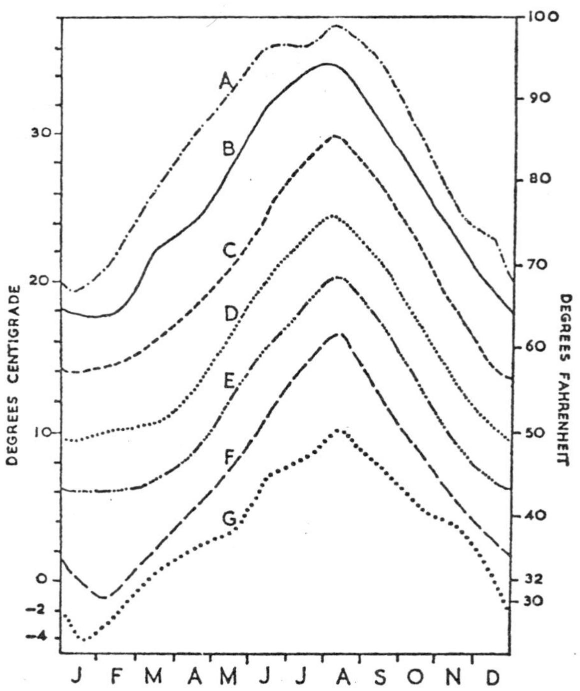

Fig. 6.6 : Monthly temperature figures for Ajaccio, Corsica.

TEMPERATURE FACTORS

Insolation

The main source of heat affecting the atmosphere and the earth's surface is the radiant energy (*solar radiation*) which travels through space from the sun (a mass of intensely hot gases, with a temperature estimated to be at its surface about 6000°C), in the form of electro-magnetic energy. The earth receives only a very small part of the total radiation, perhaps one two-thousand-millionth, but this is vital for the earth. This solar or radiant energy is known as insolation. At the outer limit of the atmosphere the solar radiation consists of visible light rays (about 41 per cent of the total), very short gamma rays, alpha rays, X-rays and ultraviolet rays (9 per cent), and the longer infrared and heat rays (50 per cent).

The measure of the intensity of insolation per unit area received at the outer limit of the atmosphere is called the *Solar Constant*, which is calculable. This is equal to nearly 2 gramme-calories per square cm per minute; *i.e.*, the radiation received each minute on a surface area of a square cm could raise the temperature of 1 gramme of water through 2°C (or alternatively equivalent to 1.4 kilowatts per sq m). In point of fact, this term seems to be a misnomer, since radiation does seem to vary, as evidenced by sun-spot activity, though only by about 1-2 per cent. Moreover, the earth has an elliptical orbit around the sun, so that it is at its farthest point on July 4 (a position known as aphelion, at a distance of 152 million km (94.5 million miles), and at its nearest on 3 January (*perihelion*, at a distance of 147 million km [91-5 million miles)]. The amount of solar energy received is therefore about 6 per cent more at the perihelion, and it would seem that as this occurs in the northern hemisphere winter, that season should be slightly milder than in the southern hemisphere. Actually, other factors, notably the greater speed of the earth along its orbit in December-January due to the curvature of its ellipse, and the great area of land in the northern hemisphere, completely mask this effect.

What really matters is the effect on the insolation when it enters the earth's atmosphere and in part passes to the earth's surface. Some 19 per cent is directly absorbed by the atmosphere, particularly by the carbon dioxide and water-vapour present; this proportion is so low because the short light-waves pass, with little check, through the atmosphere. A small amount (9 per cent) is lost as a result of diffuse reflection ('scattering') caused by air molecules and dust particles; this scattered radiation is not converted into heat. The scattering effect is greatest for the rays of light at the short-wave end of the spectrum, the blue and violet end, which helps to explain the colour of the sky. A small amount, by contrast, is down-scattered towards the earth's surface. A serious loss of insolation (25 per cent) is involved in that amount reflected directly back into space from both clouds and the ground; the radiation coefficient of the latter is termed its albedo. The drop in temperature experienced when a cloud passes across the sun can be very appreciable, and a thick 'overcast' can locally and temporarily reduce incoming insolation by as much as 80 per cent. On an average, however, about 53 per cent of incoming insolation is lost.

The balance (about 47 per cent) reaches the earth's surface to be converted into heat energy. As the surface warms, it re-radiates some of this energy, but in the form of long-wave heat radiation, equivalent to 14 per cent of the total insolation initially received. Clearly if this re-radiation did not occur, the surface would become progressively hotter. These long waves are more readily absorbed by the atmosphere than are the incoming short waves, and so the atmosphere is indirectly heated. It is also warmed by the transfer of latent heat as a result of evaporation and condensation equivalent to about 23 per cent of the total energy involved. Finally, the heated surface of the earth warms the layer of air resting upon it by direct conduction (10 per cent), which may continue as a vertical heat transference within the atmosphere known as convection', this is manifest as a rising air current, updraught, convection cell or popularly 'thermal'.

The surface of the earth therefore both absorbs and gives out radiant energy at the same time, although in the form of energy of different wavelengths; as a result the heating depends on the balance between the two (Fig. 6.7). During a clear, sunny day in summer there is a net gain and the temperature rises, reaching a maximum in the early afternoon. During the night, incoming insolation ceases, while the outgoing radiation continues. There is, therefore, a loss of heat during the night, and the coldest period is just after dawn.

The atmosphere acts rather like a greenhouse roof, since long-wave radiation cannot pass out through it as readily as incoming short-wave insolation. Something like seven-tenths of the outgoing radiation is absorbed by the atmosphere, otherwise temperatures would drop more markedly during the night and in winter. This insulating effect is increased by a cloud-cover. Hard frosts in winter occur during a clear, starlit night, when the outgoing radiation into space is at its maximum. Similarly, in the tropical deserts with their cloudless skies, terrestrial radiation at night causes a sharp fall in temperature. In California and elsewhere 'smudging' is used to combat nocturnal cooling; fires giving off dense palls of smoke check radiation and so reduce frost incidence at blossom-time.

Figures represent approximate mean percentages of the incoming isolation over the earth as a whole. Solid lines indicate short-wave radiation, pecked lines long-wave terrestrial radiation. These are average figures, which vary seasonally and in various climatic regions.

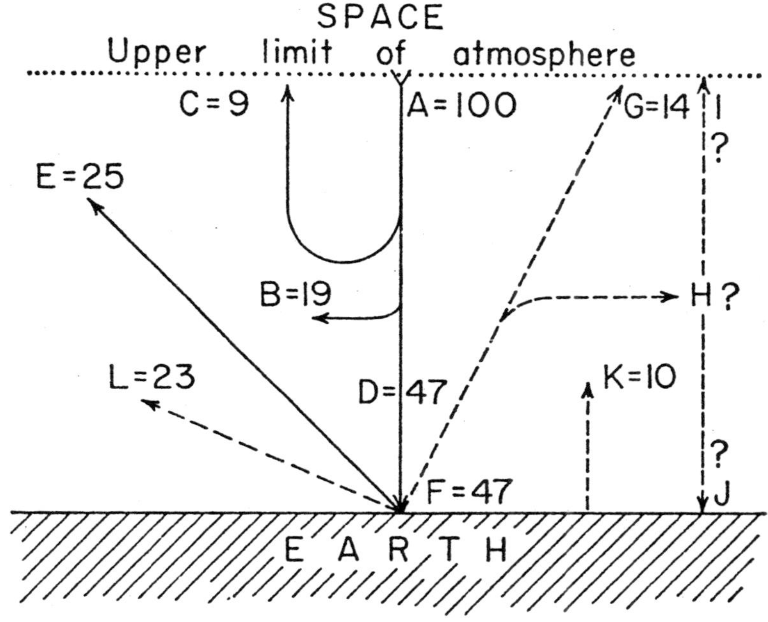

Fig. 6.7 : The mean balance of solar radiation.

Latitude

So far insolation has been discussed in general terms, without considering how it may vary for astronomical reasons. This variation is a function of latitude.

The earth both *rotates* on its own axis, which is responsible for the alternation of day and night, and revolves in its elliptical orbit around the sun for a period of about 365¼ days, which is responsible for the seasons. The axis of the earth about which it rotates is tilted at an angle of 66½° to the plane of the path it traces during its revolution (known as the plane of the ecliptic). As a result, the position of the overhead midday sun varies throughout the year; the sun appears to 'decline south'

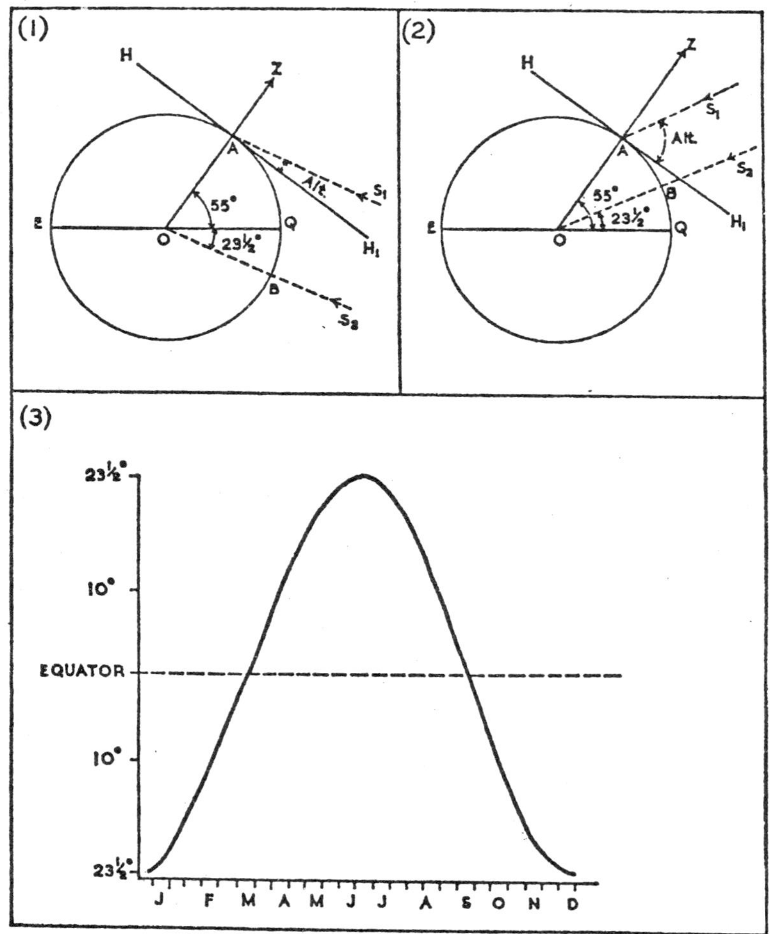

Fig. 3.9 : The altitude of the sun.

in the northern winter, and to return north in summer. The results of this change in declination are shown in Fig. 6.8. The North Pole on 21 June is tilted towards the sun (the June

Solstice), on 22 December it is tilted away from it (the December Solstice), and on 23 September and 21 March the two poles are just reached by the sun's rays (the *Equinoxes*).

These astronomical facts have two important results upon the degree of insolation. They cause very marked differences (a) in the angle of incidence of the sun's rays, and (b) in the length of day and night during the several seasons at various parts of the earth :

(a) The angle of incidence at which the sun's rays fall upon the surface of the earth influences their heating effect. Not only is the length of their path through the atmosphere increased in high latitudes (if the length of the path through the atmosphere is doubled, the amount of heating per unit of area is decreased to one quarter), but the amount of solar energy per unit of area is decreased as shown in Fig. 6.9. Difference in the angle of incidence therefore causes a proportional difference in the intensity of insolation.

(b) At the Equinoxes every parallel is half in daylight, half in darkness, and day and night are almost equal throughout the world. On 21 June, however, there is constant darkness within the Antarctic Circle; the length of day increases gradually to almost exactly 12 hours at the Equator, and at the Arctic Circle the sun does not sink below the horizon. These facts are summarized in the following table, derived from the *Nautical Almanac*:

	Diurnal duration of Daylight (in terms of monthly mean hours)			
	Equator	*50°N*	*60°N*	*70°N*
Mar.	12-13	11-81	11-70	9-55
June	12-13	16-22	18-59	No night
Sept.	12-13	12-60	12-86	15-50
Dec.	12-13	8-09	5-93	No day

The effect of the seasons is therefore to modify the duration of insolation, and the net result is that both intensity and duration of insolation must be taken into account. At the Equator the value of the insolation varies little; the day is about 12 hours long and the sun at midday is never more than 23°¼ from overhead. There are two slight maxima of insolation at the Equinoxes, two slight minima at the Solstices. With increasing distance from the Equator, the longer are the summer days; during the Solstice at the Tropic the day is of about 13½ hours and the midday sun is overhead, so that insolation is greater than it ever is at the Equator. Beyond the Tropic the length of the summer day continues to increase with latitude, Jaut the angle of incidence of the sun's rays decreases. For a while the former more than compensates for the latter, and the amount of insolation increases to a maximum at 43¼° latitude. But poleward of this the increasing length of the day fails to counterbalance the diminishing angle of the midday sun, and the value of the insolation therefore decreases.

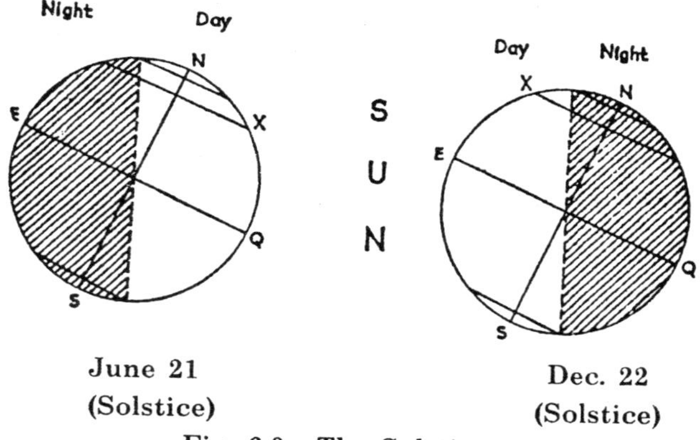

June 21 (Solstice)

Dec. 22 (Solstice)

Fig. 6.9 : The Solstices.

When all the calculations of the theoretical value of insolation are completed and the effective incoming insolation is balanced against the outgoing terrestrial radiation, there is a net heat loss in latitudes poleward of 40°, a net heat gain between the 40°N and S parallels. As the earth's surface as a whole seems to be maintained at a more or less constant temperature, the total gain and loss must balance. In other words, there must be a *heat transference* from tropical to polar regions, by the movement of both air and water-masses. This is the basis of atmospheric movements.

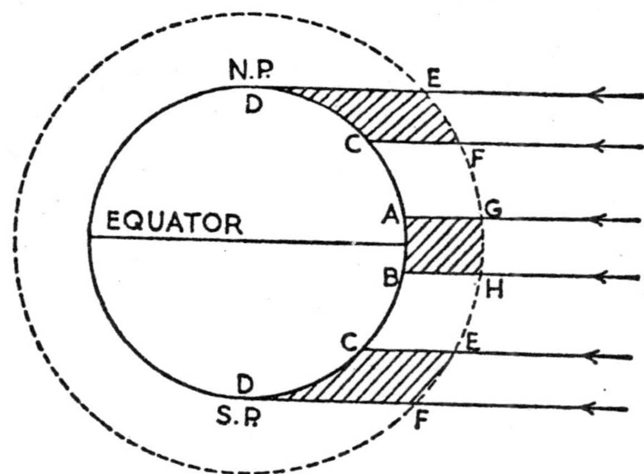

Fig. 6.10 : The angle of incidence of the sun's rays.

Twilight Particles of dust and water-vapour in the earth's atmosphere reflect the sun's light before it has appeared above the horizon in the morning and after it has disappeared below it in the evening. Astronomical twilight lasts from when the sun is 18° below the horizon until dawn, and again from sunset until the sun is 18° below the horizon. The more limited civil twilight extends only from 6° and to 6° respectively. In high latitudes the low slant of the sun's path as it dips or rises causes twilight to be prolonged as compared with low latitudes. At the North Pole twilight persists continuously from 23 September to 14 November, and then again from 29 January to 21 March, that is, during the time when the sun's declination is changing from 18° to 0° and back again; in between is a period of complete darkness.

The Nature of the Surface

So far the value of insolation has been discussed as if it were affecting a homogeneous surface. The basic difference as far as the world is concerned is between land and water surfaces.

The specific heat of a substance is defined as the number of calories required to raise the temperature of 1 gramme of that substance through 1°C; a calorie represents the amount of heat required to raise the temperature of 1 gramme of water through 1°C. On this basis the specific heat of seawater is 0.94 and of granite about 0.2; in other words, water must absorb nearly five times as much heat in order to increase in temperature by the same amount as earth. Land surfaces therefore heat up more rapidly and intensely than do water surfaces. On the other hand, land cools more rapidly when the source of heating is cut off.

Other factors emphasize this contrast. Solid earth is a bad conductor, hence a shallow layer is more intensely heated. The earth is opaque, while water allows the rays to penetrate to a greater depth and so affect more water to a lesser extent. The daily fluctuations in temperature may be detected 15-18 m (50-60 ft) below the surface of water, but not more than a metre below the surface of the ground. A water surface allows evaporation, which is a cooling process, and there is also a considerable heat transference through mixing by convection currents readily set up in water. It reflects a high proportion of the light rays (compare burnished silver, which remains cool under the hottest sun because it reflects 95 per cent of the rays), while darker earth substances absorb more. A piece of black paper left on the surface of a snowfield will in a few hours have sunk to a depth of several cm because it has absorbed heat and so melted the snow around. A snowfield reflects a large proportion of the rays, so that the snow remains while skiers become bronzed in the brilliant sun-shine; the polar ice-caps are virtually perfect reflectors. The ratio between the total solar radiation falling upon a surface and the amount reflected, expressed as a decimal or a percentage, is known as the reflection coefficient or albedo. The earth's average albedo as seen from space (including also the *albedo* of the clouds) is about 0.4; *i.e.*, four-tenths of the solar radiation is reflected back into space. Surface albedo varies from 0.03 for dark soil to 0.8 for a snowfield. Water has a low albedo (0.02) with near vertical rays, though a high one for low-angle slanting rays, and grass has about 0.25. Dry sandy soils have a low specific heat and warm up rapidly at the surface, although, the insulating effect of the air contained between the particles causes the heat to be retained at the surface, so that the sand only a few cm below may be as much as 15°C cooler. Swamps or waterlogged soils act like a water surface, and forests modify heating by casting shade.

Microclimatology presents many examples of these variations; compare these records, taken on

Salisbury Plain in June 1925, of air temperatures just above various surfaces:

	Temperature				
	°C	°F		°C	°F
Shade (air)	22	71	Bare earth	35.5	96
Macadam road	43	109	Grassland	29	85
Sand	35	95	Brick rubble	31	88

The climatological results of differential heating of land and water result in both diurnal and seasonal contrasts. The decrease in temperature during a summer night as a result of outgoing radiation is only one-quarter as great in a given volume of water as in the same volume of earth. Consequently a water surface is subject to a much smaller diurnal range.

Thermals

When heating of the earth's surface is localized as a 'hot spot' (as over a road, patch of Tarmac or bare earth, as compared with a thick vegetation cover), the rising updraught is known as a thermal or 'convection cell', which both birds and glider-pilots may use to assist their ascent. As a thermal develops, it seems to take the form of a vortex ring, with convergent air rotation within it. Rarely are the hot spot and resulting rising air column continuous; usually thermals comprise individual bodies or 'bubbles' of turbulent ascending air. There may be, it is true, some steady thermal sources on warm, windless days of anticyclonic conditions, but more usually there is a general turbulent zone in the air near the ground from which thermals are 'triggered off' at intervals.

Distance from the Sea

One of the most fundamental concepts in climatology is the distribution of seasonal temperatures over land and sea. The continents tend to be warmer in summer than oceans in the same latitudes, but appreciably colder in winter; the larger the landmass, the greater the contrast (Fig. 6.11). Thus, extreme temperatures, with a large seasonal range, are the result of continentality, which obviously affects the northern hemisphere much more than the southern. On the other hand, the oceans and those land margins affected by oceanic influences tend to have more equable temperatures, with low seasonal ranges. In the discussion of climatic types, continental and marine factors are major criteria.

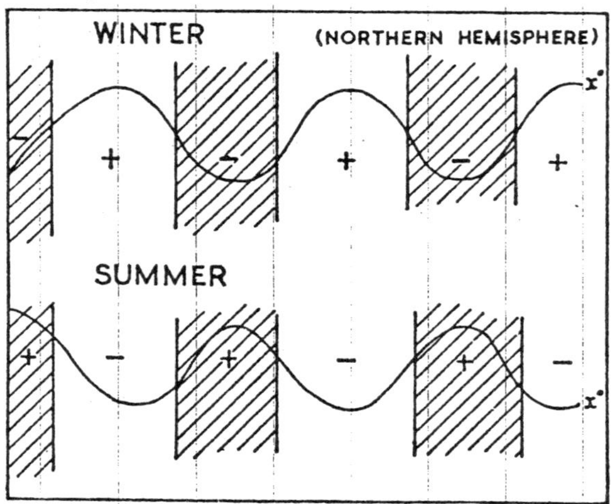

Fig. : 6.11 : The effects of land masses and oceans temperature The shaded area represents land, the unshaded area water.

Large lakes have the same general effect as an ocean, but on a less marked scale, depending on the extent of the water area. Microclimatologists have shown that the lakes of Switzerland affect temperatures a km or so from their shores. The Great Lakes of North America cause a marked northward deviation of the winter isotherms. A strip of country 50 km (30 miles) wide along the eastern shore of Lake Michigan is known as the 'Fruit Belt', and the southern shores of Lake Erie form the 'Grape Belt' of Pennsylvania and New York State; such fruit as grapes and peaches can be successfully ripened because of the warmth retained by the water-body in autumn. Paradoxically, the lake-waters warm up less rapidly in spring than the land, causing a slight chilling of the coastal belts, and premature blossoming of the fruit-trees is avoided.

Relief

One immediate relief influence on temperature has been mentioned—the effect of altitude. The environmental lapse-rate is about 0-6°C per 100 m (3.3°F per 1000 ft), but there are considerable variations from this figure. During the period 1884-1903, when an observatory was maintained on the summit of Ben Nevis at 1343 m (4406 ft), the mean annual temperature was –0.25°C, compared with 8°C (47°F) at Fort William near sea-level. In the British Isles the lapse-rate averages 0.6°C per 133 m (1°F per 410 ft) in winter, but increases to 0-6°C per 90m (1°F per 270 ft) in summer.

It must of course be remembered that there may be a considerable difference between a 'free-air' lapse-rate and a 'relief lapse-rate. Both tend to decrease by night and increase by day. Much more important variations are found when vertical movements of air occur. However, the environmental lapse-rate is of practical importance. Thus, Delhi, in the Indo-Gangetic plain at a height of 219 m (718 ft), has a mean June temperature of 33°C (92°F), but at its hill-station, Simla (2204 m, 7232 ft), the corresponding temperature is only 19°C (67°F).

Other relief controls of temperature involve shelter and aspect. Shelter protects an area from cold winds; compare, for example. the bleak cold which may be experienced at the mouth of the Rhone when the mistral is blowing, with the mild Riviera coast to the east sheltered by the *Maritime Alps*.

On a larger scale, southern China experiences abnormally low winter temperatures because it is exposed to the cold out blowing monsoon from which India is protected by the wall of the Himalayas. America has no real transverse barrier, so that New Orleans may experience 'cold snaps' brought by air streams from the Arctic, while very warm humid airstreams from the Gulf of Mexico may bring heat-waves to the northern states and even into Canada.

Aspect, including both the direction and degree of slope, is important, since a southern aspect in the northern hemisphere means greater insolation. A striking contrast in temperature, vegetation and human use is seen on the shady and sunny sides of an Alpine valley.

Inversion

One striking exception to the decrease of temperature with altitude is where an inversion occurs. During clear, settled weather, as radiation of heat takes place during the night, the air on hill-slopes is rapidly cooled, and this dense air drains downward, filling valleys or basins with cold air, possibly at temperatures below freezing, when the upper slopes are markedly warmer. Miles City, Montana, at a height of 703 m (2371 ft) in a deep hollow in the Great Plains, has experienced the lowest temperature ever recorded in the U.S.A. (excluding Alaska), –54°C (–65°F). Yet the summit of Pike's Peak (4300 m, 14,109 ft) has never recorded a temperature below –40°G (–40°F).

Frost may cause serious losses to fruit-cultivators if it occurs during the critical blossom-time; the plum-growers of the Vale of Evesham, the viticulturalists of the Gironde and the Champagne districts, and the citrus-fruit growers of California alike face the problem. Smudging with dense clouds of smoke, carefully maintained oil-heaters among the trees, and even flooding with water to check the rate of cooling and to stimulate light fogs, are used.

In southern France, where early vegetables (primeurs) are grown, the fields are covered before sundown with light straw mats, though not to exclude frost so much as to keep in ground warmth. The best method is to avoid 'frost pockets' by siting orchards on slopes away from lines of cold air drainage.

Inversions also occur as the result of the relative position of air-masses of different temperatures associated with local pressure systems. In addition to what might be called 'surface' or 'ground' inversions, similar phenomena sometimes occur at high altitudes. Their importance is that they tend to prevent vertical movement of air, hence inducing a state of stability, and so there is unlikely to be any rainfall.

The Isothermal Layer

Under very stable high pressure conditions, the temperature may remain the same within a layer of air extending to a considerable height.

Winds

The effect of winds on temperature is to 'transport' temperatures prevailing in the areas over which they blow, either from]and or sea. The result of exposure to, or shelter from, cold out blowing winds from continental interiors has been mentioned. Onshore winds in the Tropics, blowing from over the cooler ocean, tend to modify temperatures on the coastal margins. On the other hand, onshore winds such as the Westerlies may in winter carry mild temperatures from over the oceans on to the continental margins. Local winds (the warm sirocco, the cold mistral and bora, and the foehn or chinook winds on the lee side of mountains) may produce rapid temperature changes.

Ocean Currents

Ocean currents share with winds the ability to 'transport' temperatures. Where onshore winds blow over these currents, they convey similar temperature conditions to the land margins.

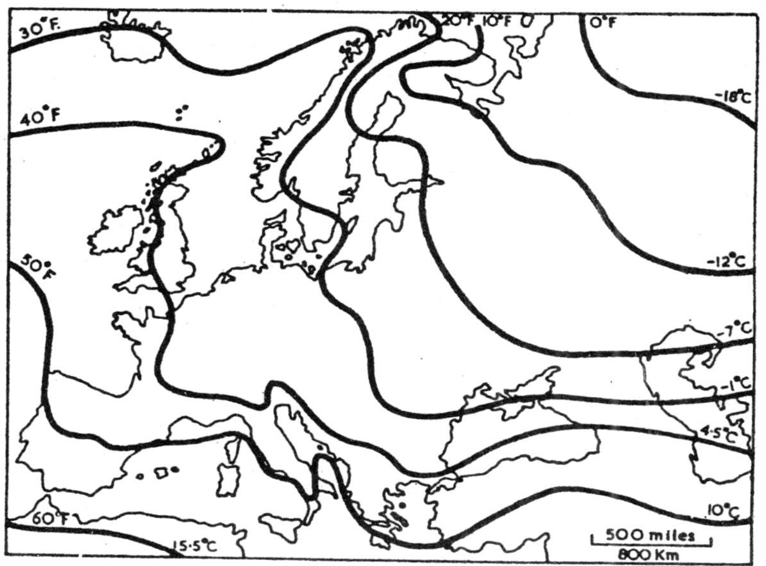

Fig. 6.12 : January isotherms/or Europe.

Especially important are the 'cold-water coasts' off the western sides of the continents in the Tropics and Sub-tropics. The 'winter gulf of warmth' which affects the North Atlantic (Fig. 6.12 on next page) is largely the result of the North Atlantic Drift; northern Norway is ice-free, though in the same latitude as central Greenland.

ISANOMALOUS TEMPERATURE

The difference of observed temperature of a place and the mean temperature of the latitude passing through that place is called *thermal anomaly*. For example, if the average temperature of 30°N latitude is 20°C and the temperature of 'A' place located on that latitude is 30°C, then the thermal anomaly is of 10°C. If the observed temperature of a particular place is more than the mean temperature of the latitude of that place, the thermal anomaly is called *positive thermal anomaly* but if the observed temperature of a given place is less than that of the latitude of that place then it becomes *negative thermal anomaly*. The equal thermal anomaly of several places is called *isanomalous temperature* and the lines drawn on the world map joining places of equal thermal anomaly are called *isanomals*.

7

AIR MASSES

MEANING AND CHARACTERISTICS

"An air mass may be defined as a large body of air whose physical properties, especially temperature, moisture content, and lapse rate, are more or less uniform horizontally for hundreds of kilometres" (R.G. Barry and R.J. Chorley, 1968). According to A.N. Strahler and A.H. Strahler (1978) "a body of air in which the upward gradients of temperature and moisture are fairly uniform over a large area is known as an air mass." An air mass may be so extensive that it may cover a large portion of a continent and it may be so thick in vertical dimension that it may vertically extend through the troposphere. It may be pointed out that since a single air mass is so large that it may cover hundreds of thousands to millions of square kilometres of the earth's surface, and hence horizontal homogeneity of an air mass in terms of its physical properties may not be practically possible because the nature and degree of uniformity of air mass properties are determined by:

(i) The properties of the source area and the direction of its movement,

(ii) Changes introduced in the air mass during its journey away from the source area, and

(iii) The age of the air mass.

The vertical distribution of temperature in an air mass, and moisture content of the air are two basic properties of an air mass which control the weather conditions of the area affected by that air mass. An air mass is designated as cold air mass when its temperature is lower than the underlying surface while an air mass is termed warm air mass when its temperature is higher than the underlying surface. The boundary between two different air masses is called front. The physical properties of an air mass is determined on the basis of the characteristic features of the surface through which it travels. An air mass also affects and modifies temperature and moisture conditions of the areas visited by it and in turn it is also modified by the local conditions of the visited areas. The concept of the air-mass is perhaps the most fundamental in modern meteorology. An air-mass consists of a large body of air with its temperature and moisture content horizontally uniform over a considerable distance. If in a vertical cross-section of such an airmass the isobars and isotherms (that is, the horizontal surfaces of constant pressure and temperature) are parallel, the atmosphere is said to be *barotropic*, and tends to be stable in character.

Air-masses can be divided into groups according to their source regions, trajectory and characteristics, both of temperature and humidity. On a basis of temperature they are known as Polar or Tropical, on a basis of humidity as Maritime (having crossed the oceans, and so moist) or Continental (originating over the continents, and so dry). Their combinations allow four main categories to be distinguished—Polar Maritime (Pm), Polar Continental (Pc), Tropical Maritime (Tm) and Tropical Continental (Tc). The name Equatorial Maritime (Em) is sometimes applied to the very moist and warm air-masses originating within a few degrees of the Equator; these are a pronounced form of Tm air. (American usage reverses these symbols, as cP and mT.)

The Pc air originates over the continental interiors, the northern tundra lands of North America and Asia, and the Greenland icecaps, forming masses of cold, dry air. As these masses spread outward toward the coasts and over the North Atlantic, they are warmed from below, and their water-content increases. Thus Pm air is formed. From the sub-tropical high-pressure cells come the Tc airmasses, moving either equatorward or poleward. In the former case the air-mass is known as TcK (German kalt = cold), because it tends to be cooler, especially in winter, than the equatorial air with which it comes in contact. When it moves poleward, it is TcW (warm).

The major air-masses from tropical latitudes are of Tm quality. They too may move towards either cooler or warmer latitudes, and are denoted by TmW and TmK respectively, although, paradoxically, as they diverge from the same air-cell reservoirs, TmK air is warmed on its equatorward progress and is warmer than TmW air which is cooled as it progresses poleward.

A further useful distinction is the addition of (M), where the TmW air-mass movement is monsoonal in character, signified by the abbreviation TmW(M). Polar continental air may also have a monsoonal effect (PcM)—the outblowing winter monsoons.

It may be desired to indicate that an air-mass has undergone considerable modification after it has left its source region. In that case a prefix N may be added; thus NTm. Some meteorologists go further still and add a suffix initial denoting the source region; thus NP for North Pacific, SI for South Indian Ocean, A for Australia, and so on. Others add a suffix S or U to indicate stable or unstable conditions. Special categories of air-mass may be denoted by A for those originating over the Arctic and AA over the Antarctic. There is a possible source of confusion between Polar and Arctic air-masses. The former term was coined merely to denote a temperature contrast with a Tropical air-mass. The term Arctic was introduced later, after the name Polar had become established, and not all authorities distinguish between them.

Frontal Zones

As air-masses of differing temperature and humidity move outward from the various major high-pressure cells, they come into conflict. A vertical section taken across such a zone of conflict will show that the surface of constant temperature and those of constant pressure will intersect at an angle; in other words, such an airmass has become *baroclinic*, which will result in a condition of large-scale atmospheric instability in the frontal zones. These become the scene of the formation of numerous small low–pressure systems, a process known as *frontogenesis*.

The most striking frontal zones are in the North Pacific and North Atlantic, where the Pm and Tm air meet. These are the Polar Fronts, the Pacific PF and the *Atlantic* PF respectively. The *Mediterranean* Front is a winter only frontal zone, lying along the tongue of sea separating the Tc air-mass over the Sahara from the PC air-mass over Europe. A *Secondary Polar Front* sometimes develops in the central North Pacific area, if the main Subtropical high pressure cell has divided into two. The Arctic Fronts (AF) lie around the northern margins of Eurasia and America.

The *Intertropical Convergence Zone* (ITCZ) lies more or less around the Equator, where Tm air-masses converge. The term Intertropical is better than Equatorial, as it has sometimes been called, since the movement of the Thermal Equator over the land-masses at the Solstices is quite appreciable. It is not a line of true air-mass discontinuity but a broad and complex zone, involving at least two air-mass boundaries, broadly parallel and sometimes distantly separated. Indeed, it has been shown that between the boundaries of the TmK air-masses over Asia and Australia respectively, lying some hundreds of kilometres apart, a distinct semi-permanent air-stream (referred to as the *Equatorial Westerlies*) can be discerned in the Indian Ocean, Indonesia and the eastern Pacific Ocean, and also over a small area in Ecuador and Colombia.

Other suggestions for names have been made such as the 'Equatorial Air-stream Boundaries', and the term 'Equatorial Trough' is also sometimes used, since it implies no causal connection.

Where the ITCZ is only weakly defined, the air-masses may be virtually stagnant; this is in fact the low-pressure area of the Doldrums. Over the ocean the ITCZ is a zone of convergence where the TmK air-masses (in effect, the Trade Winds) have more or less similar characteristics of temperature and humidity. Over the land- masses it may be a front in the correct sense, where a Tm air-mass comes into contact with a markedly different Tc air-mass, forming intertropical frontal waves or weak depressions. This is clearly shown, for example, in West Africa. Here Tc air from the Sahara high-pressure cell and Em air from the Gulf of Guinea come into contact; these air-streams are recognizable as the excessively dry Harmattan and the moist monsoonal winds respectively.

The importance of air-masses cannot be over-estimated; modern climatology is based on a study of their origin, dispersal and resulting contacts. Broadly, Equatorial climates are controlled by Equatorial and Tropical air-masses, mid-latitude climates by both Tropical and Polar air-masses, and high-latitude climates by both Polar and Arctic air-masses. Consider the following two examples, The North American continent is subject to different degrees, at different seasons, to;

(i) Cold, dry stable Ac or PC air from the north, in its most extreme form known as a 'polar outbreak' into tropical latitudes,

(ii) Cool, moist unstable Pm air from the north-west (Pacific) and north-east (Atlantic),

(iii) Warm, moist and fairly unstable Tm air from the south-west (Pacific),

(iv) Warm, moist and extremely unstable Tm air from the south-east (Gulf of Mexico and Atlantic) and

(v) In summer only, hot, dry and unstable Tc air forming over the south-western desert-lands. The climates of North America are the sum-total of the effects and interactions of these airmasses. On a different scale, the British Isles and western Europe, essentially zones of transition, are subject to Pm air from the west, north-west and north, Tm air from the south-west, PC air from the north-east and east, and Tc air from the south-east. Most critical is the frontal zone between the Pm and Tm air to the west.

SOURCE REGIONS

The extensive areas over which air masses originate or form are called source regions whose nature and properties largely determine the temperature and moisture characteristics of air masses. An air mass originates when atmospheric conditions remain stable and uniform over an extensive area for fairly long period so that the air lying over that area attains the temperature and moisture characteristics of the ground surface. Once Termed, an air mass is seldom stationary over the source region, rather it moves to other areas.

An ideal source region of air mass must possess the following essential conditions:

(i) There must be extensive and homogeneous earth's surface so that it may possess uniform temperature and moisture conditions. The source region should be either land surface or ocean surface because irregular topography and surface comprised of both land and water cannot have uniform temperature and moisture conditions,

(ii) There should not be convergence of air, rather there should be divergence of air flow so that the air may stay over the region for longer period of time and thus the air may attain the physical properties of the region. It is, thus, apparent that anticyclonic areas characterised by high barometric pressure and low pressure gradients are most ideal regions for the development of air masses,

(iii) Atmospheric conditions should be stable for considerable long period of time so that the air may attain the characteristics of the surface.

There are 6 major source regions of air masses on the earth's surface *e.g.*

(1) Polar oceanic areas (North Atlantic Ocean between Canada and Northern Europe, and North Pacific Ocean between Siberia and Canada-during winter season),

(2) Polar and arctic continental areas (snow-converted areas of Eurasia and North America, and Arctic region-during winter season),

(3) Tropical oceanic areas (anticyclonic areas—throughout the year),

(4) Tropical continental areas (North Africa-Sahara, Asia, Mississippi Valley zone of the USA—most developed in summers),

(5) Equatorial regions (zone located between trade winds—active throughout the year), and (6) monsoon lands of S.E. Asia.

CLASSIFICATION OF AIR MASSES

Any classification of air masses must consider the fact that all of their weather characteristics (mainly temperature, humidity and lapse rate) are properly represented and incorporated. Thus, the weather conditions of air masses at their source regions and thermodynamic and mechanical modification introduced in them during their journey away from their respective source regions must be taken into consideration while classifying them into definite categories. There are two approaches to the classification of air masses *e.g.*;

(1) Geographical classification, and

(2) Thermodynamic classification.

(1) Geographical Classification

The geographical classification of air masses is based on the characteristic features of the source regions. Trewartha has classified air masses on the basis of their geographical locations into two broad categories viz. (i) Polar air mass (P), which originates in polar areas. Arctic air masses are also included in this category, (ii) Tropical air mass (T), which originates in tropical areas. Equatorial air masses are also included in this category. These two air masses have been further divided into two types on the basis of the nature of the surface of the source regions (whether continental or oceanic areas) *e.g.* (a) continental air masses (indicated by a small letter (c), and (b) maritime air masses (indicated by a small letter m). It may be pointed out that a continental air mass gets modified and is transformed into maritime type while passing through ocean surface but maritime air mass is seldom transformed into continental type while passing through land surface. Based on above facts air masses are classified into the following four principal types according to their geographical locations-

(i) Continental polar air mass (cP)

(ii) Maritime polar air mass (mP)

(iii) Continental tropical air mass (cT)

(iv) Maritime tropical air mass (m t)

(2) Thermodynamic Modifications and Classification of Air Masses

Thermodynamic modification of all air mass involves its heating or cooling from below while passing through different surfaces away from the source region. Heating of an air mass causes decrease in the vertical stability of the atmosphere. After being originated the air masses move out of their source regions to other regions and in the process they modify the weather conditions of the areas travelled by them and in turn they also get modified by the surface conditions over which they move. The thermodynamic modifications of air masses, besides heating from below, also include evaporation of water into the air from below or into intermediate layer by precipitation from moist air aloft.

The modification of air masses depends on 4 factors *e.g.*;

(i) Initial characteristics of air mass in terms of temperature and moisture content,

(ii) Nature of land or water surface over which a particular air mass moves,

(iii) Path followed by the air mass from the source region to the affected area, and

(iv) Time taken by the air mass to reach a particular destination.

An air mass while moving over the surface whose temperature is greater than the lower layer of the moving air mass, is heated from below and becomes unstable due to resultant steepened lapse rate and upward movement of air. This mechanism causes condensation, cloud formation and precipitation if the moving air mass contains sufficient amount of moisture content. On the other hand, if the moving air mass is wanner than the surface over which it travels, it is cooled from below resulting into atmospheric stability which restricts upward movement of the air and thus there is no chance for condensation, cloud formation and precipitation. It is, thus, obvious that cold polar air masses while moving from their source regions to relatively warmer surfaces become unstable because they are wanned from below. On the other hand, warm tropical air masses, when move out of their source areas and reach colder surfaces, are cooled from below, causing atmospheric stability and dry weather.

A *warm air mass* (w) is that whose temperature is greater than the surface temperature of the region visited while if the air mass is colder than the surface temperature it is called *cold air mass* (k). It is apparent that the warmness or coldness of an air mass is determined by the temperature of the underlying surface. Air mass also undergoes thermodynamic modification when evaporation is added to it from outside.

An air mass is termed stable air mass when air descends while an air mass becomes unstable when upward movement of air is operative. Such *mechanical modifications* in an air mass are introduced due to cyclonic and anticyclonic conditions. Besides, mechanical modifications are also introduced due to:

(i) Turbulent mixing caused by eddies or convection,

(ii) Divergence and convergence of air masses and their effects on lapse rate of temperature,

(iii) Subsidence of air and lateral expansion on the ground surface (anticyclonic condition),

(iv) Lifting of air and convergence of air at the ground surface (cyclonic condition), and

(v) Advection.

Based on thermodynamic and mechanical (dynamic) modifications air masses are divided into (i) cold air mass and (ii) warm air mass, each of which is further divided into (a) stable air mass, and (b) unstable air mass.

(1) *Cold air masses* originate in the polar anarchic regions. They are characterized by the following properties in their source regions-

(i) Temperature is very low because of loss of heat through outgoing longwave terrestrial radiation.

(ii) Specific humidity is extremely low.

(iii) Stability increases and normal lapse rate of temperature is low.

Cold air masses after moving out from their source regions and reaching other areas have the following properties-

(i) The temperature of the areas where cold air masses reach starts decreasing.

(ii) The air mass is warmed from below and thus normal lapse rate increases and the air becomes unstable. This mechanism causes convective currents.

(iii) If the cold air mass lies over warm ocean surface, then its specific humidity increases and cumulo-nimbus clouds are formed.

(iv) The usual visibility in the air mass is maintained.

(v) Precipitation occurs only when the air mass lies over warm ocean surface but if it lies over warm continent, there is clear weather.

(vi) If the cold air mass lies partly over warm ocean surface and partly over adjoining cold land surface, then cyclonic conditions are induced.

Cold air masses are further divided into (a) continental cold air mass, and (b) maritime cold air mass.

(2) *Warm air mass* is that whose temperature is greater than the surface temperature of the areas over which it moves. Such air mass is cooled from below and thus its lower layer becomes stable due to which its vertical movement stops. Warm air masses generally originate in the subtropical regions characterized by anticyclonic conditions. They are further divided into (a) continental warm air mass, and (b) maritime warm air mass.

Based on thermodynamic and mechanical (dynamic) modifications and some other considerations air masses are divided into 16 types as follows-

(A) Continental Polar Air Masses (cP)	1. Continental Polar Cold Stable Air mass (cPKs)
	2. Continental Polar Cold Unstable Air Mass (cPKu)
	3. Continental Polar Warm Stable Air Mass (cPWs)
	4. Continental Polar Warm Unstable Air Mass (cpWu)
(B) Maritime Polar Air Masses (mP)	5. Maritime Polar Cold Stable Air Mass (mPKs)
	6. Maritime Polar Cold Unstable Air Mass (mPKu)
	7. Maritime Polar Warm Stable Air Mass (mPWs)
	8. Maritime Polar Warm Unstable Air Mass (mPWu))
(C) Continental Tropical Air Masses (cT)	9. Continental Tropical Cold Stable Air Mass (cTKs)
	10. Continental Tropical Cold Unstable Air Mass (cTKu)
	11. Continental Tropical Warm Stable Air Mass (cTWs)
	12. Continental Tropical Warm Unstable Air Mass (cTWu)
(D) Maritime Tropical Air Masses (mT)	13. Maritime Tropical Cold Stable Air Mass (mTKs)
	14. Maritime Tropical Cold Unstable Air Mass (mTKu)
	15. Maritime Tropical Warm Stable Air Mass (cTWs)
	16. Maritime Tropical Warm Unstable Air Mass (cTWu)

c = continental, T = tropical, m = maritime,

K = cold, W = warm,

u = unstable, s = stable

AIR MASSES OF NORTH AMERICA

(A) Wintertime Air Masses

(1) Continental Polar Air Mass (cP)

Continental polar air mass is cold (K), dry and stable (s), and originates over the snow-covered central Canada to the north of 50°–60°N latitude, and Alaska while continental arctic air masses originate over arctic basin and Greenland ice cap. These air masses move out of their source regions and enter the USA between the Rocky mountains and the Great Lakes. Extensive land surface with topographic homogeneity and slow anticyclonic circulation provide most ideal conditions for the origin and development of continental polar air mass. This air mass moves in southerly and south-easterly directions and brings extreme cold conditions. In fact, the arrival of this cP air mass produces intense cold waves in the vast area of the USA, with the result most of the places record temperature below freezing point.

After reaching the southern and south-eastern shores of the Great Lakes the continental polar cold air mass (cPK) is modified and thus becomes moist and unstable and yields heavy precipitation in the form of snowfall locally known as lake-effect snow.

The modified air mass, when moves eastward, is forced to move upward along the Appalachians, becomes unstable, clouds are formed and the western slopes of the Appalachian mountain receive heavy snowfall. It may be pointed out that so long as the cold continental polar air mass moves over the snow-covered surface, it is least modified but as it crosses middle Illinois and enters the snow-free surface it is warmed from below and thus the cold air mass is modified to warm continental polar air mass (cPW) and the stability of the air mass decreases. The continental polar warm air mass (cPW) while moving through east-central USA meets maritime tropical (mT) air mass and polar front is formed which induces cyclonic conditions and winter precipitation occurs to the east of the Rocky mountains. The cPW air masses are modified in the south-eastern USA due to (i) mechanical turbulence produced by corrugated terrains of the southern Appalachians, (ii) subsidence of air from above and resultant stability, and (iii) instability in the lower layer of the air masses caused due to addition of heat and moisture.

(2) Maritime Polar Pacific Air Masses(mP)

These air masses are called maritime polar pacific air masses because these originate in the northern parts of the North Pacific, mainly near Aleutian Islands where winter low pressure is formed.

This region is surrounded by continental polar air mass from all sides except in the south. The water surface is warmer than the air lying over it. Consequently, when the cold maritime polar air mass comes over this area it is warmed from below and thus becomes unstable. This mPKu air mass (maritime polar cold unstable air mass) picks up moisture throughout its journey south-eastward to the west coast of North America. This moist air mass gives sufficient precipitation while rising along the western slopes of the coast ranges. During summer season polar Pacific air mass becomes stable. When polar Pacific air mass reaches the Great Plains of the USA after crossing over the coastal ranges and the Rocky mountains, it undergoes the process of thermodynamic modifications and thus is transformed into cold, stable and dry continental polar air mass (cPWs) which induces anticyclonic conditions in the central states of the USA, with the result sky becomes clear and air circulation becomes slow and temperature returns to normal.

(3) Maritime Polar North Atlantic Air Mass (mP)

This airmass originates over the North Atlantic Ocean mainly in the region between Greenland, Newfoundland, and Labrador where winter temperature ranges between 5°F and 40°F while during summer season it becomes 50° to 60°F. Since the general air circulation in this region is from west to east and hence this air mass has little influence on North America. Some times cyclonic winds draw maritime polar North Atlantic air mass in the northeastern parts of the USA mainly to the east of the Appalachians and to the north of Cape Hatteras. This air mass is dry and stable in its upper layers while it is moist and unstable in the lower parts. This air mass brings in bad weather which is locally known as 'north-eastern' characterized by strong northeast cold winds, exceedingly low temperature well below freezing point, high moisture content in the air and possible precipitation mainly in solid forms such as snowfall, sleet and hailstorms.

(4) Tropical Maritime Atlantic Air Masses

These air masses (mT) originate over Gulf of Mexico, Caribbean Sea and subtropical western portion of the North Atlantic Ocean. They are hot, moist and unstable and are capable of unleasing heavy showers. These air masses affect most parts of the USA east of the Rocky mountains. Temperature ranges between 21°C and 26°C and remains almost uniform in the source regions, with the result tropical maritime Atlantic air mass becomes warm and moist air mass. It may be pointed out that it becomes very difficult to this air mass to enter the southern and central USA in winters because there is complete dominance of continental polar air mass over these areas. Whenever tropical maritime Atlantic air mass enters the USA it is cooled from below because it is much warmer than the ground surface and thus becomes dry and stable. This modified air mass is known as maritime tropical stable warm air mass (mTWs). This stable air mass is incapable for precipitation but whenever it comes in contact with continental polar air mass, the upper air instability increases and the air mass is modified to maritimetropical warm unstable air mass (mTWu). As this air mass rises along the mountain barriers it yields heavy showers. In summer the maritime tropical warm air mass coming from over Mexican Gulf brings hot and sultry weather in the southern and south-eastern states of the USA. This air mass also produces several thunderstorms.

(5) Tropical Maritime Pacific Air Masses

Pacific air mass originates over the subtropical portions of the East Pacific west of USA and Mexico mainly over the high pressure area located to the southwest of California. The air mass becomes stable because of subsidence of air from above due to anticyclonic circulation. This maritime tropical stable (mTks) air mass is dry, cold and stable near the Pacific coast of the USA. Whenever this air mass is associated with cyclonic circulation it becomes unstable and brings rains. This airmass seldom crosses the Rockies.

(B) Summertime Air Masses

(1) Polar Continental Air Mass (cP)

This air mass originates over the snow-covered central Canada and Alaska. The temperature becomes relatively higher in summers than in winters but adjoining oceanic areas have relatively low temperature. This air mass is cold, dry and unstable in the source regions but the air mass originating on cold arctic ocean during summer season is initially cold and stable. The stability of the cold and stable arctic air mass disappears when this air mass moves southward over relatively warm ground surface and hence is heated from below. The continental cold air mass (cPK.) becomes warm due to thermodynamic modification when it moves over the ocean. When this air mass moves southward from its source areas, it does not cover long distances but it extends for long distances in the east. The southward movement of continental polar air mass brings chilly weather in the eastern and central USA due to which the effects of summer heat waves are eliminated and fine weather sets in. Whenever the continental polar air mass is associated with cyclonic circulation, it produces sporadic rainfall in the north-central and eastern parts of the USA.

(2) Maritime Polar Atlantic Air Masses

These air masses originate over the area located between Cape Cod and Newfoundland. Initially, these are cold and stable. These reach as far south as northern Florida where temperature is reduced by 15° to 25°F. Low temperature, clear sky and full visibility are the weather characteristics associated with these air masses. Thus, these air masses produce fine and pleasant weather in the region extending from Newfoundland to Cape Hatteras. There is no ground fog due to dry condition.

(3) Maritime Tropical Atlantic Air Masses

These air masses originate near Barmuda where high pressure is formed. They move northwestward and control the weather conditions of vast areas of the USA east of the Rocky

Mountains during summer months. Thermally induced low pressure over southern and central USA draws maritime tropical air masses (mP) far inland but the existence of polar front in the vicinity of the Great Lakes restricts their entry to Canada. Since temperature and moisture content in the air increases considerably due to arrival of these air masses in the central and eastern USA, the weather becomes oppressive and unpleasant. As these air masses move put of their source areas and enter the USA after crossing over the Gulf of Mexico, surface temperature increases, and they are modified into maritime tropical unstable air masses (mTKu) because the heating of overlying relatively cold air mass causes atmospheric instability.

Thus, thunderstorms and cyclones are produced which yield heavy showers. As the air mass moves northward it loses its moisture content and becomes dry in the upper Mississippi valley. When these air masses move westward and rise along the Rocky Mountains they yield heavy downpour with cloud burst. Similarly, when they cross over the Applachians they give heavy showers through thunderstorms.

(4) Maritime Polar Pacific Air Masses (mP)

Maritime polar Pacific air masses originate in the area near Aleutian Islands in the north Pacific Ocean. The air mass becomes stable because of subsidence of air from above during summer season. Thus, this air mass becomes cold and stable (mPs). It may be pointed out that the continental surfaces are wanner than the water surfaces of the Pacific Ocean. Thus, the maritime air masses are wanned from below when they reach the continental areas.

This causes greater turbulence in the lower layer, marked decrease in relative humidity, disappearance of clouds and dry weather. No doubt, the temperature of the west coastal areas of the USA, mainly Californian coasts, is reduced during summer season with the arrival of these air masses. After crossing over the Rocky Mountains maritime polar Pacific air masses are modified and resemble continental polar air masses (cP) in physical characteristics.

(5) Maritime Tropical Pacific Air Masses (mT)

These air masses originate in the tropical North Pacific Ocean off the west coast of Central America. These air masses are marginalized because of the prevalence of maritime polar Pacific air masses (as referred to above) along the west coasts of North America in summers.

(6) Continental Tropical Air Mass (cT)

This air mass originates in the source region comprising Mexico, western Texas (USA) and eastern New Mexico (USA). The daytime characteristics are high temperature, significantly low humidity and scant rainfall. This air mass moves to Great Plains and causes extreme arid conditions. It produces drought conditions if it stays for a long period over an area.

AIR MASSES OF ASIA

(A) Winter Air Masses

(1) Continental Polar Air Maw

These air masses originate over extensive areas comprising Siberia and outer Mongolia having very cold ground surface. Initially, the air masses are very cold and dry in their source regions. The lower portion upto the height of one kilometre is characterized by inversion of temperature. The air masses move eastward and after covering long distances are mechanically modified as mechanical turbulence is produced when these air masses cross over the mountain barriers. This process leads to the disappearance of inversion layer resulting into increase of temperature and humidity in the lower layer. These air masses enter China through two routes viz. (i) through land surface, and (ii) through water surface. When high pressure lies over Mongolia and North China, then these air masses enter China by land route.

They are much warmer in China than in their source areas. These air masses are associated with clear sky and dry weather and cold air. When these air masses come with high velocity, they bring with them immense quantity of dust and sands and deposit them as loess. The continental polar air masses in their modified forms affect the weather conditions of most parts of Asia during winter season. These air masses do not enter the Indian subcontinent because of effective barrier of the Himalayas.

When high pressure lies over Manchuria and Japan Sea, the continental polar air masses enter China by sea route after moving over Japan Sea, and Yellow Sea and thus pick up abundant moisture. These air masses are relatively warmer and more humid than the continental polar air masses coming by land route. Until they are associated with fronts, they are characterized by clear sky and pleasant weather. The lower portion is unstable and thus they give precipitation when they ascend along the mountain barriers. The continental air masses coming through sea and land routes converge along the eastern coasts of Asia and form cyclones through frontogenesis and cause precipitation.

(2) Maritime Polar Air Masses (mP)

These air masses after originating over the Sea of Okhotsk influence only the coastal margins of Siberia, Manchuria and South Korea while the eastern coasts of Asia south of Korea are deprived of their influences because (i) the winter air circulation is off shore *i.e.*, from west to east due to which the westward advance of maritime polar air masses is blocked, and (ii) continental polar air mass while entering China through sea route attains the characteristics of maritime polar air mass (mP). These air masses (mP) also invade Japan in early summer and form fronts when they converge with overlying maritime tropical air masses and bring moist weather with overcast sky and light precipitation.

(3) Maritime Tropical Air Masses (mT)

Maritime tropical air masses do not effectively influence the weather conditions of the eastern Asia during winters because of the dominance of continental polar air masses. These air masses are experienced only upto southern China. They are warmer and more humid than all of the wintertime air masses. Unstable maritime tropical air masses are more effective in south-west Pacific Ocean and in eastern Indonesia.

(B) Summer Air Masses

(1) Continental Polar Air Masses (cP)

The source areas of polar air masses extend further northward in central Asia because of high temperature during summer season due to north-ward migration of the sun. The air becomes relatively warm but continental polar air masses do not effectively influence the weather conditions of eastern and southern Asia because maritime tropical air masses become more dominant during summer season. The continental polar air masses enter China only through sea route from Japan Sea and Yellow Sea. These air masses are colder than maritime tropical air masses. They are associated with clear weather, scant precipitation, and negligible thunderstorms. They produce cyclonic conditions whenever they converge with maritime tropical air masses.

(2) Maritime Tropical Air Masses (mT)

The weather of south and south-east Asia is largely controlled by maritime tropical air masses which are known as summer monsoons. They are warm, more humid and unstable. They yield torrential rainfall when they are forced to ascend by mountain barriers. After being originated in southern oceans they move north and north-eastward and after entering the mainland they are heated from below because of warm ground surface and hence they become unstable and convectional currents are produced. The south-west summer monsoons of Indian subcontinent are typical representatives of maritime tropical summer air masses. These air masses produce cyclonic conditions when they converge with continental polar air masses during springs in central China and during middle summer in Manchuria.

(3) Maritime Polar Air Maw (mP)

These air masses originate over Okhotsk Sea from where they move westward and influence the weather conditions of eastern Asia north of 40°N latitude. These air masses are more active during summers than during winters. They are more effective in Manchuria and east Siberia. Though these air masses extend upto southern Japan in early summer but later they are pushed northward by maritime tropical air masses.

8

CLASSIFICATION OF CLIMATES AND CLIMATIC TYPES

INTRODUCTION

World climates have been generally classified into certain types and subtypes on two bases *e.g.* (i) empirical basis and (ii) genetic basis. The empirical classification of world climates is based on statistical data of climatic elements derived through observations and experiments while genetic classification is based on the causes or genesis of climatic phenomena and climatic variations. Though several scientists have attempted to classify world climates into certain types but the schemes of Wladimir Koeppen, C.W. Thomthwaite, and G.T. Trewartha are more commonly used.

KOEPPEN'S CLASSIFICATION

The German botanist and climatologist Wladimir Koeppen presented his descriptive scheme of the classification of world climates first in 1900 based on vegetation zones of French plant physiologist Candolle presented in 1874. He revised his scheme in the year 1918 wherein he paid more attention to monthly and annual averages of temperature and precipitation and their seasonal distribution. He again modified his scheme in 1931 and 1936. Koeppen's original scheme was modified in 1953 by Geigger-Pohi and the revised scheme known as Koeppen-Geigger-Pohl's scheme of classification of world climates was published. It may be pointed out that the classification of Koeppen is more popular because it is quantitative in nature as numerical values of temperature and precipitation have been used in delineation of boundaries of different climatic types. The climates have been named on the basis of alphabets and climates have been determined on the basis of formulae and hence the classification has become difficult to memorise because each alphabet has definite and specific meaning.

Koeppen used five major vegetation zones of the world as identified by Candolle in 1874 (*e.g.* megatherms, xerophytes, mesotherms, microthenns, and hekistotherms) as the basis of classification of world climates on the belief that the distribution of natural vegetation was the best indicator of the total picture of climate of a region concerned. Based on these five vegetation zones he divided the world climates into 5 principal types and designated them by capital letters A, B, C, D, and E.

(1) A Climate	represents humid tropical climates characterized by winterless season, warm and moist conditions throughout the year and mean temperature always above 18°C.
(2) B Climate	represents dry climates where evaporation exceeds precipitation and there is constant water deficit throughout the year.
(3) C Climate	represents humid mesothermal or middle latitudes warm temperate climates having mild winters, average temperature of the coldest and warmest months between 8° and 18°, and below 18°C respectively.
(4) D Climate	includes humid microthermal or cold forest climates characterized by severe winters, average temperatures of coldest and warmest months being 3°C and above 10°C respectively.
(5) E Climate	includes polar climates characterized by summerless season, average temperature of the warmest month below 10°C.

Besides these capital letters, Koeppen has used the following small (lower) letters in his scheme for specific meaning.

f = precipitation throughout the year, average temperatures of the coldest month being more than 18°C, minimum precipitation of 6cm in every month of a year

m = monsoon climate, short dry season, average precipitation in driest month less than 6cm

w = winter dry season

S = well defined summer dry season

Koeppen has divided 5 major climatic types into 11 subtypes on the basis of seasonal regimes of precipitation and nature of aridity and coldness.

(1) Tropical Rainy Climates (A Climates)

A or tropical rainy climate is that where the temperature of the coldest month is above 18°C. On the basis of periodicity and regime of precipitation this type has been further divided into 4 types.

(i) *Af climate :* humid tropical climate, precipitation in the driest month more than 6cm, seasonal distribution of precipitation more or less uniform throughout the year, very low daily range of temperature.

(ii) *Aw climate :* tropical humid and dry climate, winter dry season (w), precipitation of at least one month less than 6cm, high temperature throughout the year.

(iii) *Am climate :* monsoon climate, one short dry season but sufficient annual precipitation and thus wet ground throughout the year, dense forest, precipitation of at least one month less than 6cm. The boundary between Aw and Am climates is demarcated on the basis of annual precipitation and the precipitation of the driest moth as per formula given below–

$$a = 3.94 - r/25$$

where a = precipitation of driest month

r = annual precipitation

If the precipitation of the driest month of a place is less than the value of a, it will be Aw climate, if it is more than the value of a, it will be Am climate.

Example = If the annual precipitation of a place is 50 inches, then Am/Aw boundary would be = 3.94 - 50/25 = 1.94 inches. If the precipitation of the driest month of that place is 2 inches (this should be always less than 2.4 inches, otherwise it would be Af climate), it would be Am climate. On the other hand, if the precipitation of the driest month of that place is 1.8 inches, it would be Aw climate.

(iv) *As climate :* dry summers, rarely found.

The aforesaid Af, Am and Aw climatic types as identified by Koeppen are generally similar to equatorial rainforest climate (Af), monsoon climate (Am), and savanna climate (Aw) respectively. Koeppen has further identified finer details in A climates and has used the following lower letters to indicate them–

w' maximum precipitation in autumn

w" two seasons of maximum precipitation separated by two dry seasons

s dry summers

i difference of temperature of the warmest and the coldest month less than 5°C.

g hottest season preceding precipitation

(2) Dry climates (B climates)

Evaporation exceeds precipitation, precipitation not sufficient to maintain permanent stable water table of ground water. B climates are divided into two types on the basis of annual temperature and the rainiest month of the year *e.g.* (i) dry desert climate (BW), and (ii) semi-arid or steppe climate (BS). The boundary between BW and BS climates is determined on the basis of the following formula–

$$r = \frac{0.44t - 8.5}{2}$$

where

r = annual precipitation (inches) .

t = temperature (0°F)

If the annual precipitation of a given place is more than the value of r, the climate of that place will be BS but if it is less then r, the climate will be BW. For example, if the temperature of a place is 80°P, the annual value of precipitation for dividing boundary between BS and BW climates will be 13.3 inches as given below–

$$r = \frac{0.44 \times 80 - 8.5}{2} = 13.3 \text{ inches}$$

B climates are further differentiated on the basis of annual temperature. When the mean annual temperature is more than 18°C (64.4°F), the climate is indicated by h letter but if the mean annual temperature is less than 18°C, it is indicated by k letter. Thus, B climates are divided into the following four types–

(i) BWh tropical desert climate, average annual temperature more than 18°C (64.4°F)

(ii) BSh tropical steppe climate, mean annual temperature above 18°C

(iii) BWk middle latitude cold desert climate, mean annual temperature below 18°C

(iv) BSk middle latitude cold steppe climate, mean annual temperature below 18°C

Koeppen has identified further details in B climates and has used the following letters to indicate them–

k mean annual temperature below 18°C

h mean annual temperature above 18°C

a summer dry, three times more precipitation in the wettest month of winter season than the driest month of summer season

w winter dry, 10 times more precipitation in the wettest month of summer season than the driest month of winter season n minimum fog

(3) *Humid mesothermal or warm temperate rainy climates (C climates)* : average temperature of the coldest month above 3°C but below 13°C, precipitation in all seasons. Based on seasonal distribution of precipitation C climates have been divided in to 3 climatic types–

(i) *Cf climate* : precipitation throughout the year, precipitation more than 1.2 inches in the driest month of summer season. This climate represents Western Europe type of climate. This is further divided into two second order sub-divisions *e.g.* Cfa (humid subtropical) and Cfb (marine west coast type)

(ii) *Cw climate* : dry winters, 10 times more precipitation in the wettest month of summer season than the driest month of winter season. This represents China type of climate.

(iii) *Ca climate* : dry summers, three times more precipitation in the wettest month of winter season than the driest month of summer season, precipitation of the driest month of summer season less than 1.2 inches. This represents Mediterranean type of climate.

Koeppen has identified further minor details in C climates and has used a few explanatory small letters as given below-

a warm summers, temperature of the warmest month above 22°C (71.6°F).

b cold winter, temperature of the warmest month below 22°C.

c cold short summer season

i, n, g = as explained above.

(4) *Humid microthermal or cold snow forest climates or humid cold climates (D climates)* : temperature of the coldest month below-3°C (26.6°F) but of the warmest month above 10°C (50°F), ground surface covered with snow for several months of a year. This climate has been divided into three types.

(i) *Df Climate :* humid cold climate, no dry season. This is further divided into (a) Dfa (long warm summers, continental), (b) Dfb (long and cool summers), and (c) Dfc (short cool summer-subarctic).

(ii) *Dw Climate :* humid cold climate, dry winters, further divided into (a) Dwa-continental climate with long cool summer, (b) Dwb-cool short summer (sub-arctic type), and (c) Dwc-cold winters. d = temperature of the coldest month below –38°C, f, a, w, b, c as explained above.

(5) *Polar climates (E climates) :* temperature of the warmest month less than 10°C (50°F), further divided into (i) ET and (ii) EF climates.

(i) *ET climate :* tundra climate, temperature of the warmest month below 10°C but above 0°C.

(ii) *EF climate :* permanent snow fields, temperature in all months below 0°C.

Evaluation Of Koeppens's Scheme

Koeppen used two easily measurable weather elements *e.g.* temperature and precipitation as the basis for statistical parameters for the delineation of different climatic regions. In fact, temperature and precipitation are most widely and most frequently used effective weather elements as representatives of the effects of climatic controls. His scheme of climatic classification is primarily based on the relationship between floral types and their characteristics, and climatic characteristics of a given place or a region. He also paid due consideration to the loss of moisture through evaporation as he included effective precipitation, which depends on the rate of potential evapotranspiration, in his scheme. It may be pointed out that it is not the total annual precipitation which matters more for vegetation community rather it is the effective precipitation (amount of precipitation which is actually available to plants) which is more important for flora. Koeppen's scheme appealed more to geographers because the scheme recognized association between vegetation types and climatic types. Besides, this scheme is descriptive, generalized and simple and hence it was widely acclaimed.

Inspite of several merits as referred to above the Koeppen's scheme also suffers from some serious drawbacks. Koeppen gave undue significance to mean monthly values of temperature and precipitation in his scheme of climatic classification and neglected other weather elements such as precipitation intensity, amount of cloudiness and number of rainy days, daily temperature extremes, winds etc. He made his scheme more descriptive and generalized and ignored the consideration of ausative factors of climate. He did not include the characteristics of different airmasses in his classification. The use of different letter symbols to indicate different climatic types and their secondary and tertiary subtypes makes the scheme very difficult to memorise.

CLASSIFICATION BY THORNTHWAITE

C. W. Thornthwaite, an American climatologist, presented his first scheme of classification of climates of North America in 1931 when he published the climatic map of North America. Later he extended his scheme of climatic classification for world climates and presented his full scheme in 1933. He further modified his scheme and presented the revised second scheme of classification of world climates in 1948. His scheme is complex and empirical in nature.

(1) 1931 Classification

Like Koeppen Thornthwaite also considered natural vegetation of a region as the indicator of climate of that region. He accepted the concept that the amount of precipitation and temperature had paramount control on vegetation but he also pleaded for inclusion of evaporation as important factor of vegetation and climate. This is why Thornthwaite used two factors, *e.g.* precipitation effectiveness and temperature effectiveness, for the delimitation of boundaries of different climatic regions.

(2) Precipitation Effectiveness

Precipitation effectiveness or precipitation efficiency refers to only that amount of total precipitation which is available for the growth of vegetation. He used precipitation efficiency ratio for the calculation of this amount of water available to vegetation. Precipitation efficiency ratio (P/E ratio) is calculated by dividing total monthly precipitation by monthly evaporation and precipitation efficiency index (P/E index) is derived by summing the precipitation efficiency ratios for 12 months of a year. Since it is difficult to obtain data of evaporation for every cenfe and hence Thornthwaite suggested the following formulae for the calculation of precipitation efficiency ratio and index–

$$P/E \text{ Ratio} = 11.5\,(r/t - 10)^{10/9}$$

$$P/E \text{ Index} = \sum_{i=1}^{12} 11.5\,(r/t - 10)1079$$

where

r = mean monthly rainfall in inches

t = mean monthly temperature in 0°F

He identified 5 humidity zones on the basis of P/E Index and boundary values for the major vegetation zones.

Humidity Zone	*Vegetation*	*P/E Index*
A (Wet)	Rainforest	127
B (Humid)	Forest	64-127
C (Subhumid	Grassland	32-63
D (Semiarid)	Steppe	16-31
E (Arid)	Desert	<16

Thomthwaite further subdivided each humidity zone into 20 subhumidity zones on the basis of seasonal distribution of precipitation.

1 Ar	5. Br	9. Cr	13. Dr	17. Er
2. As	6. Bs	10. Cs	14. Ds	18. Es
3. Aw	7. Bw	11. Cw	15. Dw	19. Ew
4. Ad	8. Bd	12. Cd	16. Dd	20.Ed

where r = adequate rainfall in all seasons

s = rainfall deficient in summer

w = rainfall deficient in winter

d = rainfall deficient in all seasons.

(ii) *Thermal, Effectiveness :* He believed that temperature had important contribution in the growth of vegetation.

He, thus, devised an index of *thermal efficiency* or *temperature effectiveness*, expressed by positive departure of monthly mean temperatures from freezing point, and suggested the following formulae–

(i) Thermal Efficiency Ratio

(T-E Ratio) = (t-32)/4

(ii) Thermal Efficiency Index

$$(T - E \text{ Index}) = \sum_{i=1}^{12} (t-32)/4$$

where t = mean monthly temperature in 0°F. It is apparent that T-E Index is the sum of thermal efficiency ratios for 12 months. On the basis of T-E index Thomthwaite divided the world into 6 temperature prov.ııces-

Temperature Province	*T-E Index*
A' - Tropical	127
B' - Mesothermal	64-127
C' - Microthermal	32-63
D' - Taiga	16-31
E' - Tundra	1.15
F' - Frost	0

Thus, on the basis of precipitation effectiveness, thermal efficiency, and seasonal distribution of rainfall there may be 120 probable combinations and hence climatic types on theoretical ground but he depicted only 32 climatic types on the world map as given below–

1. A A'r	tropical wet climate with rainfall adequate in all seasons
2. A B'r	mesothermal wet climate with adequate rainfall in all seasons
3. A C'r	microthermal wet climate with adequate rainfall in all seasons
4. B A'r	tropical humid climate with adequate rainfall in all seasons
5. B A'w	tropical humid climate with rainfall deficient in winter
6. B B'r	mesothermal humid climate with adequate rainfall in all seasons
7. B B' w	mesothermal humid climate with rainfall deficient in winter season
8. B B's	mesothermal humid climate with rainfall deficient in summer season
9. B C'r	microthermal humid climate with adequate rainfall in all seasons
10. BC's	microthermal humid climate with rainfall deficient in sum- mer season'
11. C A'r	tropical subhumid climate with adequate rainfall in all seasons
12. C A'w	tropical subhumid climate with deficient rainfall in winter seasons
13. C A'd	tropical subhumid climate with rainfall deficient in all seasons
14. C B'r	mesothermal subhumid climate with adequate rainfall in all seasons
15. C B'w	mesothermal subhumid climate with rainfall deficient in winter season
16. C B's	mesothermal subhumid climate with rainfall deficient in summer season

17. C B'd	mesothermal subhumid climate with rainfall deficient in all seasons
18. C C'r	microthermal subhumid climate with rainfall in all seasons
19. C C's	microthermal subhumid climate with rainfall deficient in summer season
20. C C'd	microthermal subhumid climate with rainfall deficient in all seasons
21. D A'w	tropical semiarid climate with deficient rainfall in winter season
22. D A'd	tropical semiarid climate with rainfall deficient in all seasons
23. D B'w	mesothermal semiarid climate with rainfall deficient in winter season
24. D B's	mesothermal semiarid climate with rainfall deficient in summer season
25. D B'd	mesothermal semiarid climate with rainfall deficient in all seasons
26. D C'd	microthermal semiarid climate with rainfall deficient in all seasons
27. E A'd	tropical arid climate with rainfall deficient in all seasons
28. E B'd	mesothermal arid climate with rainfall deficient in all seasons
29. E C'd	microthermal arid climate with rainfall deficient in all seasons
30. D'	taiga type climate
31. E'	tundra type climate
32. F'	permanently snow-covered polar climate

(2) 1948 Classification

After making sizeable modifications Thorthwaite presented his modified scheme of climatic classification in 1948. Though he again used previously devised three indices of precipitation effectiveness, thermal efficiency and seasonal distribution of precipitation in his second classification but in different way. Instead of vegetation, as done in 1931 classification, he based his new scheme of climatic classification on the concept of *potential evapotranspiration* (PE) which is in fact an index of thermal efficiency and water loss because it represents the amount of transfer of both moisture and heat to the atmosphere from soils and vegetation (evaporation of liquid or solid water, and transpiration from living plant leaves) and thus is a function of energy received from the sun. It may be pointed out that potential evapotranspiration is calculated (and not directly measured) from the mean monthly temperature (in 0°C) with corrections for day length (*i.e.,* 12hours). The PE (Potential Evapotranspiration) for a 30-day month (a day having only the length of sunshine *i.e.,* 12 hours) is calculated as follows-

PE (in cm) = 1.6(10t/I)a

where PE = Potential Evapotranspiration

I = the sum for 12 months of (t/5)'514

a = a further complex function of I

t = temperature in 0°C

Thomthwaite developed four indices to determine boundaries of different climatic types *e.g.*;

(i) moisture index (Im),

(ii) potential evapotranspiration or thermal efficiency index (PE),

(iii) aridity and humidity indices, and

(iv) index of concentration of thermal efficiency or potential evapotranspiration.

(i) *Moisture Index (Im) :* Moisture index refers to moisture deficit or surplus and is calculated according to the following formula

Im = (100S-60D)/PE

where Im = monthly moisture index

S = monthly surplus of moisture

D = monthly deficit of moisture

The sum of the 12 monthly values of Im gives the *annual moisture index.*

$$\text{Annual Moisture Index} = \sum_{i=1}^{12} (100S - 60D)/PE$$

(ii) *Thermal Efficiency Index :* Thermal efficiency is simply the potential evapotranspiration expressed in centimetres as expressed above. It is, thus, apparent that the thermal efficiency is derived from the PE value because PE in itself is a function of temperature. The method of the calculation of PE is given above.

(iii) *Aridity and Humidity Indices :* These indices are used to determine the seasonal distribution of moisture adequacy. These are calculated as fellows.

Aridity Index = in moist climates annual water deficit taken as a percentage of annual PE becomes aridity index.

Humidity Index = in dry climates annual water surplus taken as a percentage of annual PE becomes humidity index.

(iv) *Concentration of thermal efficiency* refers to the percentage of mean annual potential evapotranspiration (PE) accumulating in three summer months.

On the basis of *moisture index* (Im) Thornthwaite identified 9 moisture or humidity provinces.

Moisture Index	**Humidity province**
(1) 100 and above	A per humid
(2) 80 to 100	B_4 Humid
(3) 60 to 80	B_3 Humid
(4) 40 to 60	B_2 Humid
(5) 20 to 40	B_1 Humid
(6) 0 to 20	C_2 Moist subhumid

(7) –33.3 to 0	C_1 Dry subhumid
(8) –66.7 to –33.3	D Semiarid
(9) –100 to –66.7	E Arid

On the basis of *thermal efficiency* (potential evapotranspiration) 9 thermal provinces were recognized.

Thermal Efficiency Index (cm)	*Thermal Province (Type)*
(1) 114 and above	A' Megathermal
(2) 99.7 to 114.0	B'_4 Mesothermal
(3) 85.5 to 99.7	B'_2 Mesothermal
(4) 71.2 to 85.5	B'_2 Mesothermal
(5) 57.0 to 71.2	B'_1 Mesothermal
(6) 42.7 to 57.0	C'_2 Microthermal
(7) 28.5 to 42.7	C'_1 Microthermal
(8) 14.2 to 28.5	D' Tundra
(9) Below 14.2	E' Frost

On the basis of summer concentration of thermal efficiency the world was further divided into 8 provinces-

Summer Concentration of Thermal Efficiency (%)	Type
(1) below 48.0	a'
(2) 48.0-51.9	b'
(3) 51.9-56.3	b'_3
(4) 56.3-61.6	b'_2
(5) 61.6-68.0	b'_1
(6) 68.0-76.3	C'_2
(7) 76.3-88.0	C'_1
(8) above 88.0	d'

On the basis of seasonal moisture adequacy 2 major and 10 subclimatic types were identified–

	Moist Climates (A,B,C_2)	*Aridity Index*
(1) r	little or no water deficit	0 to 10
(2) s	moderate summer deficit	10 to 20
(3) w	moderate winter deficit	10 to 20
(4) s_2	large summer deficit	above 20
(5) w_2	large winter deficit	above 20
	Dry Climates (C_1,D,E)	*Humidity Index*
(6) d	little or no water surplus	0 to 16.7
(7) s	moderate winter surplus	16.7 to 33.3
(8) w	moderate summer surplus	16.7 to 33.3
(9) S_2	large winter surplus	above 33.3
(10) w_2	large summer surpuls	above 33.3

The climate of a place, thus, is determined by combining the aforesaid elements of the climatic classification *e.g.*, moisture index, thermal efficiency index, summer concentration of thermal efficiency, and seasonal moisture adequacy (aridity and humidity indices). Thus, the climate of a place is represented by four letters. For example– A A' a'r climate = Perhumid (A) megathermal (A') climate with summer concentration of annual thermal efficiency (PE in cm) of less than 48 per cent (a') and little or no water deficit (r) etc. On the basis of above indices the classification system becomes so complex due to large number of climatic types that it becomes difficult to represent them cartographically.

Evaluation of Thorthwaite's Schemes

In many aspects the 1931 classification scheme of Thornthwaite was almost similar to Koeppen's scheme because both had a few common points *e.g.* (i) Like Koeppen's scheme his scheme is also empirical as well as quantitative as the boundaries of different climates are determined on the basis of quantitative parameters derived from precipitation and temperature, (ii) Vegetation is made as the basis for the are used to designate different climatic types etc. The Thornthwaite's scheme differs from the Koeppen's scheme in that the former used two indices of precipitation efficiency and thermal efficiency for differentiation of different climatic types but the delimitation of climatic boundaries on the basis of these two indices becomes difficult and vague. Moreover, the Thornthwaite's scheme yielded the number of major climatic types (32) three times greater than Koeppen's climatic types. Like Koeppen's scheme Thornthwaite's scheme also became popular among zoologists, botanists and geographers but it was not appreciated by meteorologists and climatologists because this scheme did not include the causative factors of climates into the classification of world climates in different types. This scheme also suffers from a serious problem of non-availability of the data of evaporation for all the places. Thus, the lack of adequate climate data makes it difficult for the precise demarcation of climatic boundaries.

Though 1948 scheme of climatic classification of Thornthwaite was thoroughly revised and modified and was based on 4 important indices of moisture index, thermal efficiency or potential evapotranspiration index, seasonal moisture adequacy (aridity and humidity indices), and summer concentration of thermal efficiency but no world map of different climatic types could be prepared. It may be pointed out that it becomes very difficult to cartographically represent a large number of climatic types identified quantitatively on the basis of aforesaid indices. Moreover, the data of potential evapotranspiration are not available for all the places for a worldwide classification of climates. The complex empirical formulae divided by Thornthwaite require regular data but these are not always forthcoming. They also involve a lot of calculations for determining the climatic type of a particular place. This is why his scheme could not get more popularity and recognition.

CLASSIFICATION OF G.T. TREWARTHA

G.T. Trewartha, an American climatologist, made several revisions and modifications in the scheme of climatic classification of Koeppen since 1930s and ultimately presented his simple scheme of climatic classification having a blending of both empirical and genetic schemes of classification of world climates. In fact, Trewartha's basic aim was to present a simple, generalized and unambiguous scheme of the classification of world climates so that the major climatic types at world level could be easily and realistically identified and cartographically represented on world map. Thus, Trewartha's scheme is a compromise between purely empirical and genetic methods of climatic classification. He was fully convinced that the schemes should not be cumbersome and complex as were the schemes of Koeppen and Thornthwaite. He was also opposed to produce large number of climatic types on the basis of statistical and quantitative parameters. That is why he recognized only a limited number of major climatic types. According to him, if required, several second and third-order subdivisions may be added within each major climatic type. Like other scientists he also made precipitation and temperature as the basis for his scheme of climatic classification. He identified 6 major climatic types of first order at world level and designated them as A, B, C, D, E, F climates out of which B climates were determined on precipitation criteria while others were determined on the basis of temperature criteria.

(1) Tropical Humid Climates (A Climates)

A climates or tropical humid climates are found in those low latitudes on either side of the equator which are characterized by high temperature and adequate rainfall throughout the year and absence of winter season. On the basis of variations in precipitation A climates are subdivided into (i) Af, (ii) Aw, and (iii) Am climates,

(i) Af climate is tropical wet climate which extends upto 5° to 10° latitudes on either side of the equator and is characterized by adequate rainfall throughout the year. This is also known as tropical rainforest climate, There is no winter season as it is characterized by uniformly high temperature all the year round.

(ii) Aw climate is a tropical wet and dry climate characterized by uniformly high temperature throughout the year but there are more than two dry months. This climate is also known as savanna climate which is dominated by dry trade winds or subtropical anticyclones during winter season and by equatorial westerlies and intertropical convergence during summer season.

(iii) Am climate is monsoon climate which receives more than 80 per cent of annual rainfall during four summer monsoon months.

(2) Dry Climates (B Climates)

The boundaries of B (dry) climates have been determined on the basis of precipitation variations. They extend from the outer boundary of A climates to the middle latitudes. B climates are characterized by high evaporation, loss of moisture through evapotranspiration exceeding the annual receipt of water gain from precipitation, large annual and daily ranges of temperature, extreme seasonal temperatures, very low and highly variable annual precipitation, extremely low relative humidity, abundant sunshine and clear sky.

On the basis of aridity and annual average precipitation B (dry) climates have been divided into two climatic types *e.g.* (i) arid or desert climate– BW climate, and (ii) semi-arid or steppe climate– BS climate. On the basis of temperature variations arid (BW) and semiarid (BS) climates have been divided into 4 climatic types as follows-

(i) BWh climate	tropical-subtropical hot desert climate
(ii) BWk climate	middle latitudes or temperate and boreal cold dry climate
(iii) BSh	climate tropical-subtropical steppe or semiarid climate
(iv) BSk climate	middle latitudes or temperate and boreal steppe climate

The boundary between hot dry and cold dry climates is determined on the basis of 32°F (0°C) isotherm of the coldest month. Tropical-subtropical dry (BWh) and steppe (BSh) climates are dominated by dry trade winds and subtropical anticyclones resulting into constant dry conditions. At least 8 months of a year ‿ecord average temperature above 10°C. On the other hand, temperate cold dry (BWk) and cold steppe (BSk) climates are located on the leeward sides of the mountains in the interior of the continents and are dominated by cold anticyclones during winter season.

(3) Middle Latitudes Wet Climates (C Climates)

The isotherm of 18°C of the coldest month forms the equatorward boundary of C climates. On the basis of seasonal distribution of precipitation C climates are divided into 3 types *e.g.* (i) Cs climate (subtropical subhumid climate with dry summer, also known as Mediterranean climate), (ii) Ca climate (subtropical humid climate), and (iii) Cb climate (middle latitude marine climate). Cs climate, located on the western sides of the continents on the tropical margins of the middle latitudes, is affected by subtropical anticyclonic conditions in summers and by wet westerlies in winters. Ca climate (Cfw of Koeppen), located on the eastern sides of the continents, receives precipitation in all seasons but summer months receive more rainfall than winter months (this climate is known as China type of climate). Cb climate is affected by westerlies throughout the year.

(4) Microthermal or Temperate Climates (D Climates)

These climates are found in the areas of high middle latitudes which are affected by westerlies in summers and by polar winds in winters. The poleward and equatorward boundaries are determined by average temperatures of 10°C for 4 months in the case of the former and for 6 months in the case of the latter. On the basis of temperature variations 'D' climates have been divided into 4 types *e.g.* (i) Da climate (continental humid climate with temperature of the wannest month above 25°C), (ii) Db climate (continental humid climate with temperature of the warmest month below 22°C), (iii) DC climate (subpolar climate, short summer season), and (iv) Dd climate (temperature of the coldest month less than –38°C).

(5) Boreal Climate (E Climate)

Boreal climate is located in the higher middle latitudes and is characterized by short and cool summer season, long and very cold winter season, very short frost free season, one to three months of a year having average temperature of 10°C or more etc.

(6) Polar Climate (F Climate)

Summer season is absent. Polar winds dominate throughout the year. These climates are found in the northern hemisphere only. No month of the year records average temperature above 10°C. On the basis of temperature variations 'F' climates are divided into (i) tundra climate (Ft climate) and (ii) icecap climate (Ff climate).

Departures From Koeppen's Classification

The Trewartha's scheme of climatic classification registers the following departures from the scheme of Koeppen.

(1) In B climates Koeppen used isotherm of 18°C average annual temperature to differentiate the boundary between hot dry and cold dry (h/k boundary) climates while Trewartha used an isotherm of 32°F (0°C) of the coldest month for the determination of h/k boundary.

(2) Koeppen used the isotherm of –3°C (26.6°F) temperature of the coldest month for determining boundary between B and C climates while Trewartha selected isotherm of 32°F (0°C) for the purpose.

(3) Koeppen divided C climates on the basis of seasonal distribution of precipitation into 3 types *e.g.* (i) Cs (summers dry), (ii) Cw (winter dry), and (iii) Cf (no dry season) but Trewartha divided C climates into (i) Cs, (ii) Ca, and (iii) Cb types.

(4) Koeppen divided D climates on the basis of precipitation into (i) Dw and (ii) Df types while Trewartha divided them on the basis of summer temperature into (i) Da, (ii) Db, and (iii) Dd types.

Evaluation

As stated earlier Trewartha's scheme of climatic classification is very simple, unambiguous and a mixture of both empirical and genetic methods of climatic classification.

It uses only two weather elements *i.e.*, precipitation and temperature and avoids vigorous statistical and mathematical calculations in determining climatic type of a place and demarcating boundaries between two different climatic types. This scheme also includes the effects of land and water surfaces on the climate of an area. Trewartha's scheme became more popular among geographers because of its simplicity.

EQUATORIAL CLIMATE OR TROPICAL RAIN FOREST CLIMATE (AF)

Location

Equatorial type of climate, also known as tropical rainforest wet climate or simply Af climate, is located upto 5° to 10° latitudes on either side of the equator (Fig. 8.1) but at some places it extends upto 15°-25° latitudes mainly along the eastern margins of the continents. This climatic zone is subjected to seasonal shifting due to seasonal shifting of pressure and wind belts consequent upon the northward and southward migration of the sun. The equatorial climate is characterized by two major properties *e.g.* (i) uniformly high temperature throughout the year, and (ii) uniformly adequate rainfall throughout the year received through convective mechanism. The equatorial climate is found in the following localities—(i) the Amazon Basin in South America, (ii) the Congo Basin in Africa, (iii) Guinea coast in Africa, (iv) much of the Indo-Malaysian Region mainly in Java, Sumatra, Borneo, Malaysia, Singapore and New Guinea, (v) Philippine Islands, (vi) eastern central America (parts of Panama, Costarica, Nicargua, Honduras, Guatemala etc.), some islands in the Caribbean Sea, western Columbia and eastern Madagascar.

Temperature

Since mid day sun is almost overhead throughout the year and there is little difference between the lengths of day and night during the year and hence the equatorial region receives maximum amount of insolation which causes uniformly high temperature throughout the year as the average monthly temperature is always more than 18°C. The mean monthly temperature of most of the places ranges between 24°C and

27°C. Mean annual temperature is around 20°C but the maximum temperature of the year touches 30°C. The mean annual range of temperature of island areas ranges between 0.5°C and 1°C but other areas record annual ranges of temperature between 2°C to 3°C. The annual range of temperature of liquids (located in Peru falling in the Amazon Basin, 4°S of the equator) is 2°C. Similarly, Akassa (located at the mouth of the Niger River, Africa) records annual range of temperature of 2°C but Para records less than 2°C as annual range of temperature. The annual range of temperature becomes minimum over the oceans. For example, Jaluit, located on Marshall Island in the central Pacific Ocean records annual range of temperature of only 0.4°C. Thus, uniformly high temperatures of the equatorial regions, though lower than the temperatures of the hot desert climate, becomes unpleasant and injurious to human beings because of its uniformity and monotony.

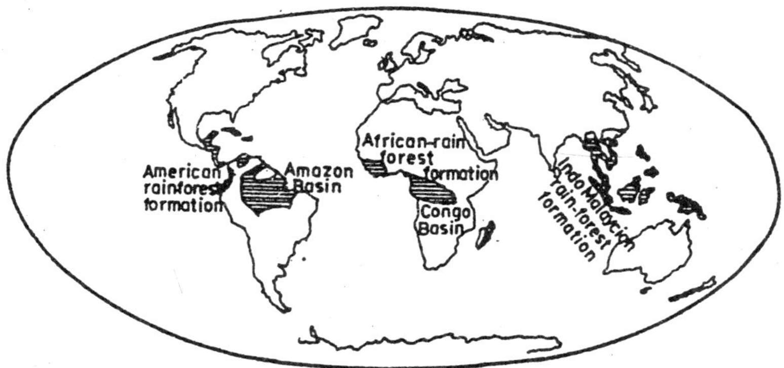

Fig. 8.1 : Equatorial rainforest (Af) Climate.

The daily range of temperature varying between 5°C and 10°C is usually far greater than the annual range of temperature. Usually, mid-day temperature rises to 290C-340C and comes down to 210C-240C during nights. Thus, the relatively low nocturnal temperature becomes uncomfortable to local people. This is why nights of the equatorial regions are called winters of the tropics. The annual range of temperature of Bolobo of the Belgian Congo is 1°C but the daily range becomes 9°C. Belam city records daily maximum and minimum temperatures as 32.8°C and 20.1°C respectively thus registering diurnal range of 12.7°C. Similarly, Santaram of the Amazon Basin records 35.5°C and 19.5°C as maximum .and minimum daily temperatures and thus diurnal range of temperature becomes 16.0°C. The daytime temperature becomes oppressive and unbearable due to high relative humidity, weak air circulation, bright sunlight etc. The inhabitants of the equatorial regions are so used and habitual to uniformly high temperatures throughout the year that they feel immediately even a slight fall in temperature. They feel cold if temperature falls below 20°C and they burn wood to ward off relative cold (though there is no winter season).

Air Pressure And Winds

Thermally induced low pressure belt develops around the equator due to uniformly high temperature throughout the year but the pressure gradient is so low that strong air circulation is not possible. Thus, the equatorial region lies in the belt of calm and doldrums characterized by light and variable winds. The surface air is heated, becomes light and moves upward thus forming convective currents. On the ground surface the winds uniformly spread laterally due to more or less uniform air pressure. The discontinuous belt of doldrum characterized by equatorial westerlies is found along the equator. The convergence of trade winds coming from the subtropical high pressure belt forms inter tropical convergence (ITC) which is associated with atmospheric disturbances (cyclones). The winds become strong with thundestonns. Strong winds give temporary relief from sultry weather. Temperature is lowered due to arrival of Harmattan winds in the nights in the Guinea coast and thus the pleasantly cooling effect of Harmattan gives comfort to human bodies. Sea breezes penetrate upto 48-96 km inland in the coastal areas and thus brings pleasant weather through their cooling effects. This is why coastal areas in the equatorial regions are comparatively better suited for living than the interior areas.

Precipitation

Equatorial regions receive rainfall throughout the year and thus there is no dry season. Average annual rainfall exceeds

200cm to 250cm. Even the driest month of the year receives rainfall more than 6cm. Iquitos (Peru), Akassa and Ocean Island receive 261 cm, 366cm and 213cm of rain per annum respectively. Though most of the rainfall occurs through convective mechanism but wherever mountain barrier becomes effective the amount of rainfall increases substantially. For example, annual rainfall reaches 1000cm in the foothill zone of Cameroon Mountain in Africa. Most of the annual rainfall in the equatorial region is received in the form of convectional rainfall (see types of rainfall in chapter 36). The strong daily vertical convective mechanism due to intense heating of ground surface because of high amount of insolation, horizontal convergence of trade winds forming intertropical convergence, a fairly large number of atmospheric disturbances (cyclonic storms) and thunderstorms yield heavy rainfall daily throughout the equatorial regions.

Distribution-inspite of high rainfall throughout the year there is no uniform spatial distribution of rainfall in all parts of equatorial climatic region. Though no month goes dry but definitely some months of the year receive more rainfall than the other months. Thus, the months having more rainfall are called wet months while the months receiving less rainfall are known as less wet months. If the temporal distribution of rainfall in the equatorial regions is considered carefully it appears that there are two periods of maximum rainfall and two periods of minimum rainfall in a year. Normally, April and November receive maximum rainfall but the period of maximum rainfall varies spatially. For example, Lagos receives maximum rainfall in May, June and July amounting to 100cm whereas October records 25cm of rainfall.

Cloudiness-equatorial climate is characterized by fairly large amount of cloudiness throughout the year. Generally, cumulus type of clouds dominates daily weather conditions. On an average, there is about 60 per cent cloudiness daily. The maximum amount of cloudiness is found between 3 and 4 P.M. daily because of maximum convective activity during this period but the sky is generally clear in the morning and at night. Though the daily period of cloudiness is less in comparison to high middle latitude areas dominated by temperate cyclones but there is strong heavy downpour due to convective mechanism and resultant convectional rainfall.

Rainfall regime : equatorial rainfall is convectional in character wherein there is daily heavy downpour from cumulo-nimbus clouds. The sky is usually free from clouds in the early morning. As the sun rises above the horizon, the amount of insolation received at the ground surface increases and hence air temperature also increases accordingly. Air is heated, becomes light and moves upward and thus becomes unstable which causes convectional system. The ascending air cools at the dry adiabatic lapse rate (10°C per 1000m) and the air soon becomes saturated and condensation level is reached. Clouds are formed. In the beginning they are cumulus and few in number but as the day advances, humidity increases due to increasing evaporation, the clouds are thickened and darkness increases. By afternoon the whole sky becomes overcast with thick cumulo-nimbus clouds. Thus, heavy rain starts with lightning and cloud thunder. As the day draws towards evening the rains become slow and weak and they completely stop by evening, clouds are cleared and weather becomes pleasant for some time. The aforesaid mechanism is repeated daily. Some times, this rhythemic daily mechanism is interrupted as rains continue uninterruptedly for several days. Continuous rains for 40 hours have been reported in Ivory Coast (Africa). The most characteristic feature of equatorial rainfall is that it is usually associated with strong thunderstorms as about 75 to 150 rainy days are associated with them.

Variability of rainfall : inspite of very high mean annual rainfall there is temporal variability in the amount of rainfall and this variability is more than the variability of temperature. The crops grown in this climatic region are such that they require more moisture and thus the year having little less than average annual rainfall is termed as a drought year because the crops are damaged. Though the word drought is unfamiliar in equatorial climatic region but some times brief drought conditions are created. There is also variability of rainfall within a year. For example, Belem (in the Amazon Basin) receives average annual rainfall of 239 cm and March receives more rainfall than other months. March has 28 rainy days against

10 rainy days in November. It may be concluded that rainfall in equatorial rainforest climate is adequate enough to support field crops and luxurious dense forests. Most of the rains is intercepted by forest canopy and thus reaches ground surface slowly in the form of aerial streamlets through leaves, branches and stems of trees and thus there is maximum infiltration of rainwater. Recent clearance of rainforests in the equatorial regions in general and in Amazonia in particular has converted once forest-converted surface into a bare ground surface which is subjected to accelerated rate of soil erosion due to daily heavy rains.

Effects of Climate On Natural Vegetation

The tropical rainforest or equatorial climatic region accounts for the largest number of plant species and luxuriant growth of natural vegetation due to high temperature and high rainfall throughout the year. The climatic region is characterized by broad-leaf evergreen dense forests comprising valuable trees such as mahogany, rosewood, coconut palm, avony, cincona, plaintain, bamboos, wild rubber, sandal wood etc. The number of tree species is so large and their diversity is so great that one hectare of land in the equatorial region accounts for 40 to 100 species. It may be pointed out that tree species account for 70 per cent of the total plant species of the tropical evergreen forests. Creepers or climbers are the second important members of the rainforests. The creepers comprising (i) climbers of lower strata, (ii) long woody climbers known as 'lianas', and (iii) epiphytes are so circuitous and highly irregular in form that it becomes difficult to find out their actual length. "They ramble through the forest, scaling the highest emergent trees and frequently looping down to the ground and then ascending further sections of the forest" (Furley and Newey, 1983). The climbers so greatly bind several trees and plants together that the accessibility in the forest cover becomes almost zero.

The vertical stratification of vegetation community consists of 5 layers or strata viz. (i) *first* or *top layer* or *dominant layer* representing the canopy of tallest trees (30 to 60m in height), (ii) *second layer* or *co-dominant layer* (25m to 30m in height), (iii) *third layer* of smaller trees (12m to 20m in height),

(vi) *fourth layer* of shrubs (5m in height), and (v) *ground layer* of herbaceous plants and ferns.

MONSOON CLIMATE (AM)

Location

Monsoon climate is generally related to those areas which register complete, seasonal reversal of wind direction and are associated with tropical deciduous forests but there are some departures from this close relationship and near correspondence between the regions of monsoon climate and tropical deciduous forests. Monsoon climate is found in the zone extending between 5° and 30° latitudes on either side of the equator (Fig. 8.2). In fact, this zone comes under the domain of trade wind belt which experiences seasonal shifting due to northward and southward migration of the sun. Onshore winds blow for six months from warm tropical oceans towards the continents and offshore winds blow for another six months from land to the sea. The areas of monsoon climate are divided in the following categories.

(1) *True monsoon areas* include India, Burma (Myanmar), Pakistan, Bangladesh, Thailand, Combodia, Laos, North and South Vietnam, southern China, Philippines, and northern coastal area of Australia.

(2) *Areas of monsoonal tendencies or pseudomonsoons* are found along the south-west coast of Africa including the coasts of Guinea, Sierra Leone,

Liberia and Ivory Coast; eastern Africa and western Madagascar (Malagasy).

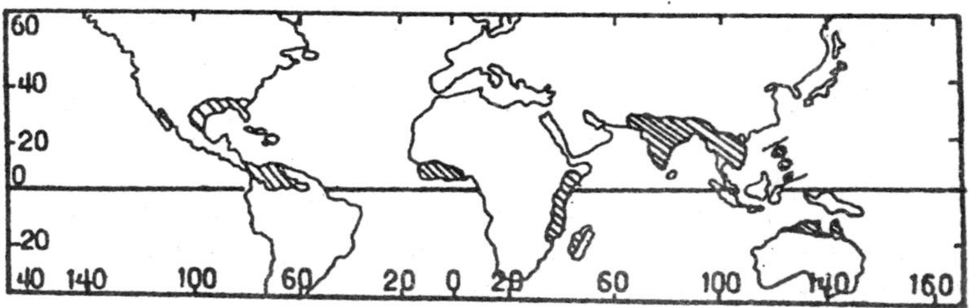

Fig. 8.2 : Distribution of monsoon climate areas.

(3) *Areas of monsoon effects* include northeast coast of Latin America *e.g.* east Venezuela, Guyana, Surinam, French Guyana, and north-east Brazil. Besides, Puerto Rico and Dominican Republic in the Caribbean Island also enjoy mild monsoonal effect.

(4) Areas of modified monsoon are found in parts of central America and south-east USA.

Temperature

Though mean annual temperature is fairly high but summer and winter seasons are sharply differentiated due to northward (summer solstice) and southward movement of the sun (winter solstice). There are three main seasons in a year in Indian Subcontinent and surrounding monsoonal areas *e.g.* (1) dry warm summer season (March to June), (2) humid warm summer season (July to October), and (3) dry winter season (November to February). Average temperature of warm dry summer months ranges between 27°C and 32°C but maximum temperature ranges between 38°C and 48°C during May and June. Warm humid summer months record average temperatures ranging between 20°C and 30°C.

The mean temperature during day in winter months varies from 10°C to 27°C. Annual range of temperature ranges between 2°C and 11°C and is controlled by nearness or remoteness of the sea (*i.e.*, distance from the sea), continental, latitudinal and altitudinal influences. For example, annual range of temperature increases inland (5.3°C at Rangoon, 11°C at Mumbai, 18.4°C at Allahabad, 20.2°C at Agra). Similarly, diurnal range of temperature is low in the coastal areas but it increases inland.

Diurnal range of temperature is much higher in dry summer season than in other seasons. For example, in the Ganga plains of India maximum temperature during day time may go as high as 44°C to 48°C and the minimum temperature during nights may come down as low as 20°C to 25°C thus registering diurnal range of 23°C to 24°C. The temperature during May and June becomes exceptionally high due to prevalence of hot winds locally known as loo.

Air Pressure and Winds

Monsoon areas are affected by high and low pressure systems due to winter and summer seasons respectively. In fact, there is complete reversal of pressure gradients over Asiatic landmass because of northward and southward migration of the sun and consequent differential heating of the continent and adjoining oceanic areas. Due to southward migration of the sun after 23 September (autumnal equinox) high pressure centres are developed on the landmass of Asia during winter season while low pressure is developed in the southern Indian and Pacific Oceans, with the result pressure gradient is developed from land areas to the oceanic areas resulting into the outflow of surface winds from high pressure centres of the land areas towards oceanic low pressure areas. This wind system having northeast direction is called winter monsoon which is nothing more than the reestablished north-east trade winds which are displaced during summer season due to northward shifting of intertropical convergence (ITC) because of northward migration of the sun. These offshore winds are generally dry because they come from over the land areas but wherever they pass over the oceanic areas, they pick up moisture and yield rainfall when effectively obstructed. For example, north-east monsoon winds while passing over Bay of Bangal pick up moisture and give rainfall in the coastal regions of Tamil Nadu during winter season. The pressure system is completely reversed during summer season when the sun registers northward migration after 21 March and becomes almost vertical over tropic of Cancer on June 21. Thus, thermally induced low pressure develops due to very high temperature over huge landmass of Asia. These low pressure centres are further intensified due to northward movement of intertropical convergence (ITC) upto 20° to 35°N latitutdes while high pressure centres are developed over southern Indian and Pacific Oceans, with the result sea to land pressure gradient is steepened and onshore winds are generated. These onshore winds are the south-west or summer monsoon winds which are moist as they pass over the sea surfaces.

According to the advocates of the concept of thermal origin of monsoon the north-east or winter monsoons and south-west or summer monsoons are originated due to differential heating of landmasses and oceanic areas during summer and winter seasons and resultant thermally induced high and low pressure areas. According to them the summer and winter monsoons are south-east and north-east trade winds. In fact, the south-east trades while crossing over the equator during summer season due to northward shifting of pressure belts caused by northward migration of the sun becomes south-westerly in direction according to Ferrel's law. Since these winds come from over the ocean, they become moist and give rainfall. According to the advocates of the dynamic origin of monsoon the belt of doldrums and North Intertropical Convergence (NITC) are drawn over south and south-east Asia during summer season due to northward migration of the sun and thus equatorial westerlies of the belt of doldrum are also established over south and south-east Asia and thus they become south-west monsoons. The tropical disturbances (cyclones) associated with the intertropical convergence (ITC) yield copious rainfall. For detailed discussion on the origin of monsoon see chapter 35.

Precipitation

Monsoon regions receive most of their annual rainfall through cyclonic and orographic types of rains though convective mechanism also yields some rainfall. On an average, the average annual rainfall is around 1500mm but there are much variations in the temporal and spatial distribution. Some times, a few areas receive less than 500mm of mean annual rainfall. Even the temporal distribution of rainfall within a single year is highly variable because more than 80 per cent of mean annual rainfall is received within 3 wet months of summer season (July, August, and September). Thus, the rainy season records much surplus water whereas dry winter and summer seasons have marked deficit water because dry seasons receive less than 25 mm of rainfall per month. There is maximum evaporation during warm dry summer months which results in desiccation of soils and marked reduction in soil water. This seasonal regime of annual rainfall gives deciduous character to the vegetations which shed their leaves during the transitional period between winter and summer seasons.

Most of the annual monsoonal rainfall is received through moisture laden south-west monsoon winds which come from over the ocean surface. The outbreak of monsoon generally occurs in India around mid-June. When these moisture laden monsoon winds strike the mountain barriers they give copious rains, This is why the western coastal plains of India receive more than 2500mm of annual rainfall because the Arabian Sea Branch of monsoon winds are obstructed by the Western Ghats and hence they are forced to ascend and soon become unstable and saturated. The leeward sides of the mountains fall in rainshadow region because descending winds are adiabatically wanned and thus become stable and dry. This is why Mangalore, located on the windward side of the Western Ghats receives 2000mm of annual rainfall whereas Bangalore, located on the leeward side receives only 500mm of annual rainfall.

The eastern coast of Tamil Nadu and Andhra Pradesh receives much rainfall during winter season through north-east monsoons as they while passing over the Bay of Bengal, pick up sufficient moisture and yield rainfall. January and February are generally driest months in India but the Ganga plains receive some rains from westerly disturbances or temperate cyclones coming from Mediterranean Sea.

Variability of rainfall in terms of both amount and duration is the characteristic feature of monsoon climate. Secondly, monsoonal rainfall is basically cyclonic in character.

Natural Vegetation

The number of plant species is far less in the monsoon climatic regions than the equatorial climatic regions. The height of most of the trees ranges between 12m and 30m. There are four strata or layers in vertical structure of the tropical deciduous forests. The uppermost and second strata consist of trees, the third stratum is formed by shrubs whereas the ground stratum represents herbaceous plants. Most of the trees are deciduous but the shrubs of the third stratum are evergreen. The trees are characterized by thick girth of stems, thick, rough and coarse bark and large hydromorphic leaves or small, hard xeromorphic leaves. Deciduous trees of monsoon climate denote complete adaptibility to wet-dry climate. The large hydromorphic leaves enable the trees to trap more and more rainfall during wet season but these are shed in dry periods to conserve moisture while small and hard xeromorphic leaves enable the trees to withstand dry weather and water deficiency. Important species of trees include sal, teak, bamboo, mango tree, mahua, jamun, neem, shisham etc.

The tropical and subtropical monsoon deciduous forest biome is one of the most disturbed ecosystems of the world. The forests have been so rapidly destroyed through both natural (forest fires) and anthropogenic processes due to rapacious utilization of forest resources for commercial and industrial purposes and large-scale clearance through mass felling of trees for agricultural land that the vegetation cover has shrunk to a very critical size. More than 80 per cent of annual rainfall occurring in only 3 wet months of July, August and September through heavy rainstorms generates maximum surface run- off which is causing enormous loss of rich fertile soils through accelerated rate of soil erosion.

SAVANNA CLIMATE (AW)

Location

The word savanna has been used for different meanings by various scientists *e.g.* the word savanna region has been used by the climatologists to indicate a particular type of climate *i.e.*, tropical wet-dry climate (Aw climate of Koeppen) as savanna climate, while the botanists have used the word savanna for a typical type of vegetation community of tropical regions characterized by the dominance of grasses. This climate is also called as Sudan type of climate. Savanna type of climate is located between 5°-20° latitudes on either side of the equator (Fig. 8.3). Thus, savanna climate is located between equatorial type of climate (Af) and semi-arid and subtropical humid climate. In other words, this climate is located between equatorial low pressure belt or rain producing intertropical convergence and subtropical high pressure belt. The most characteristic areas of savanna climate include the Llanos of Orinico Valley including Columbia and Venezuela, the Guiana Highlands, the Campos of Brazil (south central parts), and Paraguay in South America;

hilly areas of Central America; southern gart of Zaire, Angola, Zambia, Mozambique, Tanjania, Uganda, and Central Rhodesia, all to the south of the Congo Basin, and central Nigeria, southern Kenya and Uganda, Central African Republic, Dahomey, Togo, Chad, Ghana, Ivory Coast and eastern Guinea in Africa; northern Australia and some areas of India (the savanna of India is not the original and natural vegetation cover rather it has developed due to human interference with the original forest cover resulting into the development of widespread man-induced grasslands).

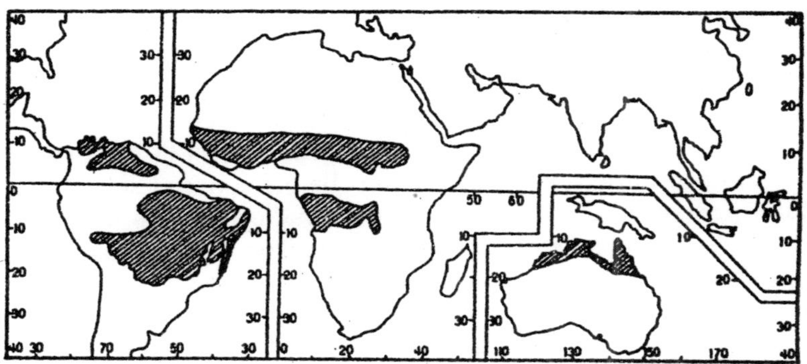

Fig. 8.3 : Location of Savanna climate.

Temperature

The Savanna climate is characterized by distinct wet and dry seasons, mean high temperature throughout the year (ranging between 24°C and 27°C), and abundant insolation. Temperature does not fall below 20°C in any month of the year. Thus, Savanna climate is similar to equatorial climate as regards temperature but the annual range of temperature ranging between 3°C and 8°C is greater than in the equatorial climate. There are three main seasons on the basis of the combination of temperature and humidity (though on an average there are only two seasons as referred to in the beginning but the dry season is further divided into warm dry season and cold dry season on the basis of temperature). (1) Cold dry season is characterized by high day temperature ranging between 26°C and 32°C but relatively low temperature during nights, usually

21°C. (2) Warm dry season is characterized by almost vertical sun's rays, high temperature ranging between 32°C and 38°C due to abundant insolation. (3) Warm wet season receives between 80 to 90 per cent of the total annual rainfall and thus records relatively lower temperature than warm dry season.

Air Pressure and Winds

The regions of Savanna climate are affected by low and high pressure systems in a year. Due to northward migration of the sun during summer solstice (21 June) the equatorial low pressure belt and doldurm are shifted northward and thus Savanna climate comes under the influence of *Inter Tropical Convergence* (ITC) which is associated with atmospheric disturbances (cyclones) which yield rains. Due to southward migration of the sun during winter solstice (23 December) Savannaclimatic zone comes under the influence of subtropical high pressure belt and thus anticyclonic conditions dominate the weather and bring dry conditions. The descending stable winds under anticyclonic conditions cause dry conditions. Besides, the coastal areas are affected by local winds and sea breezes. Eastern coasts are influenced by trade winds. Strong and high velocity tropical cyclones dominate the weather conditions during warm season. It is apparent that the Savanna type of climate is induced due to the introduction of wet summer and dry winter seasons because of northward and southward migration of the sun respectively. Since the Savanna climate is located between equatorial wet and tropical dry climates and hence there is gradual variation in weather conditions away from the equator as the aridity increases poleward.

Rainfall

The average annual rainfall ranges between 100cm and 150cm but there is much pronounced variation in the spatial distribution of mean annual rainfall in different parts of Savanna climate mainly because of two factors viz. (1) distance from the equator, and (2) the nature of topographic features. For example, the Savanna region of Brazil, locally called as Cerrado having the average absolute relief of 1300m AMSL, records mean annual temperature and mean annul rainfall of 200C-260C and

75cm-200cm respectively. The Llanos of Columbia is characterized by mean annual rainfall of 200cm-400cm (near Andes Mountain), mean annual temperature of 22°C, and maximum temperature of 32°C. The Indian Savanna is characterized by the highest temperature (being 450C-480C. in May and June) and lowest temperature (being 5°C or even less during the month of January) of all the Savanna regions of the world and the mean annual rainfall well below 150cm, 80 to 90 per cent of which is received during a brief period of 3 months (July to September).

Since the Savanna climate is a transitional belt between humid equatorial and dry tropical climates and hence there is much variability in the amount and duration of rainfall at the wet (equatorward) and dry (poleward) boundaries. There is copious rainfall in the equatorward margins because of convergence of surface winds and convective mechanism of ascending unstable winds but at the poleward margins near dry climate (BWh) is found and rainfall significantly decreases due to descending air and anticyclonic conditions.

Here mean annual rainfall becomes as low as 25 cm. As the Intertropical Convergence (ITC) moves northward due to northward migration of the sun, thunderstorms begin to develop by March and the amount of rainfall increases in the following months.

The Intertropical Convergence reaches its northernmost limit by August. Thus, rainfall continues to increase, upto August and most of rains are received through tropical cyclones and thunderstorms. Rainfall decreases due to southward shifting of ITC and dry trade winds are re-established after November resulting into dry weather condition.

It is apparent that the period of wet weather decreases while that of dry weather increases as the distance from the outer limit of the equatorial climate increases. The amount of mean annual rainfall also decreases from east to west. Savanna climate is also characterized by high variability of rainfall as there may be so heavy and abundant rainfall in a particular year that floods are caused while the following year may receive so little rainfall that drought conditions prevail.

Vegetation

Though general characteristics of typical Savanna vegetation are trees and grasses but the Savanna Biome is, no doubt, dominated by grasses. The Savanna vegetation community has developed layered structure wherein three distinct layers have clearly developed.

(1) *The ground layer* is dominated by various types of grasses and herbaceous plants, The grasses are generally coarse, stiff and hard having the height ranging between 80cm to 350cm. The African elephant grass attains enormous height of 500cm. The grasses bear deserted look during dry summer season but they become lush green again during humid summer season.

(2) *The middle layer* consists of shrubs and stunted woody plants.

(3) *The canopy layer* is formed by trees of various types. The general characteristics of trees depend on the availability of water and moisture and therefore there is great taxonomic variety of Savanna trees which are usually 6-12m in height. The Savanna trees have developed various unique characteristics to withstand dry conditions. For example, there are a few species of trees which have developed such mechanisms which help them to reduce evapotranspiration from their leaves during warm dry season and enable them to remain green even during dry season of deficient water supply.

On the other hand, there are such tree species which cannot withstand dry conditions and therefore they shed their leaves and bear the characteristics of deciduous trees. The roots of Savanna trees have also developed according to the environmental conditions as they are very large which can penetrate into the spils and ground upto the depths from 5m to 20m so that they can obtain water from groundwater even during dry season when the groundwater table falls considerably.

The small plants and many herbaceous plants have special kinds of root systems characterized by root tubers and swellings so that they may preserve water which may be used by plants during dry season. On the basis of the proportion of fees and grassland and the structure of vegetation, the Savanna vegetation is divided into 4 types.

(1) *Woodland savanna* is dominated by trees and shrubs and absence of epiphytes but some climbers having their roots in the ground are found.

(2) *Tree savanna* represents relatively open vegetation cover wherein trees and shrubs are sparsely distributed.

(3) *Shrub savanna* is represented by treeless vegetation which is dominated by grasses at the ground layer and shrubs at the second layer,

(4) *Grass savanna* is characterized by general absence of trees and shrubs and overdominance of dense grasses.

Frequent fires, both natural and anthropogenic (deliberate annual burning of grasses by man), are common features of all the savanna biomes. The rapidly increasing human population for the last 50 years or so has put enormous strain on the natural savanna grasslands because a vast area of the original grasslands has been converted into agricultural fields to grow more food crops to feed the teeming millions. Thus, the areas of natural savanna grasslands have shrunk considerably. The savanna ecosystem, thus, has been greatly degraded and destabilized.

TROPICAL-SUBTROPICAL HOT DESERT CLIMATE (BWH) (Sahara Type Of Climate)

Location

The hot desert or Sahara type of climate is located between the latitudinal belt of 150-300(350) in both the hemispheres on the western parts of the continents. This climate is found in :

(1) Africa—the Namib and Kalahari deserts of coastal Angola and southwest Africa, interior Botswana and South Africa, and Sahara desert;

(2) Asia—Thar deserts of India and Pakistan, Arabian deserts, Iranian desert;

(3) South America—Acatama desert of coastal Peru and Chile;

(4) Mojave and Arizona deserts of south-western USA;

(5) Australia—Great Sandy Desert, Great Victoria Desert and Tanami Desert (Fig. 8.4). This climate is characterized by annual aridity, and subsiding warm air masses of the subtropical anticyclones. The following reasons are held responsible for the genesis of perpetual aridity of the tropicalsubtropical hot desert climate–

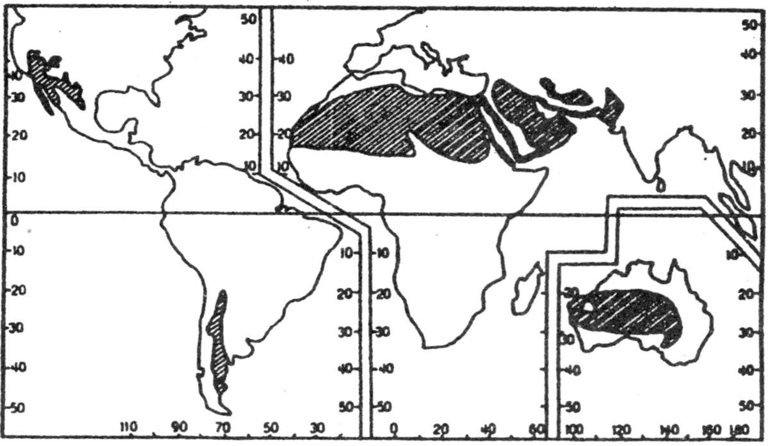

Fig. 8.4 : Location of tropical-subtropical hot dry desert climate-Sahara type of climate.

(i) Temperate cyclones do not reach these areas,

(ii) Intertropical Convergence (ITC) also does not influence these areas because of their distant location from the equator,

(iii) The trade winds spend most of their moisture through rainfall in the eastern margins of the continents and as they reach the western margins of the continents they become dry and hence are unable to give rainfall,

(iv) Due to anticyclonic conditions winds descend from above and hence they are warmed adiabatically with the result their moisture retaining capacity increases resulting into marked decrease in relative humidity,

(v) Subtropical high pressure system causes divergence of surface winds which is antagonistic to rainfall,

(vi) The ground temperature is so high that raindrops, if formed at all, are evaporated before they reach the ground surface.

Temperature

On the basis of annual distribution of temperature two distinct seasons are recognized *e.g.* summer season and winter season. Average temperature during summer season ranges between 30°C and 35°C but maximum temperature exceeds 40°C during mid-day. The temperature of 40°C to 48°C is very common at noon during summer months. The western part of Great Australian Desert records temperature above 40°C for 64 days in continuation and above 32°C for 150 days in continuation but temperature falls at nights giving much relief to the people. Phoenix of Arizona (USA) records more than 32°C at noon but temperature falls to 24°C in the nights. Azizia has recorded the highest temperature of 58°C (136.4°F) so far. Similarly, exceptionally very high temperature of 56.4°C (134°F) has been recorded in the Death Valley of Californian desert (USA). Day time mean temperature during winter season ranges between 15.5°C and 21°C but some times it reaches 27°C but at nights temperature falls to 10°C.

It is, thus, apparent that both annual and diurnal ranges of temperature are high in the tropical-subtropical hot desert climatic areas. Generally, annual range of temperature ranges between 17°C and 22°C while daily range varies from 22°C to 28°C. Some times, daily range of temperature exceeds 40°C. Very high daily and annual range of temperature is because of open and clear skies, vegetation-free ground surface, very low humidity, distance from the equator, dominance of sands etc. It may be pointed out that in the absence of clouds and moisture maximum insolation is received at the ground surface. Loose sands are soon heated and thus ground temperature soon shoots up. Again there is rapid loss of heat from the sandy surface through outgoing longwave terrestrial radiation at nights due to clear sky and completely dry condition (total absence of moisture in the air) resulting into considerable fall in night temperature. This mechanism causes very high daily range of temperature. It may be remembered that blankets are needed in the nights in hot desert areas due to very high difference in daytime and night temperatures even during summer seasons. Tripoli recorded highest and lowest temperatures of 91°F and 31°F respectively on December 25, thus registering a daily range of 60°F.

Pressure and Winds

Poleward areas of the regions of hot desert climate are affected by divergent air circulation and anticyclonic conditions because they fall in the belt of subtropical high pressure. The winds become stable and dry because they descend from above and are heated and thus there prevails dry condition. The north-east trades (northern hemisphere) become dry when they reach the western parts of the continents in the latitudinal zones of 15°-350. Some local low pressure centres are formed during summer season and thus a few local but weak cyclonic storms are produced. The upper air anticyclonic conditions do not allow the winds of these local storms to rise. Heat waves dominate in summer season thus making human life very difficult. The extensive deserts of Sahara and Australia become ideal source regions for the development of tropical continental airmasses.

Rainfall

Rainfall in tropical desert climate is so low and variable that it becomes difficult to determine average annual rainfall as it never comes true. The various sources put the annual average of rainfall between 25 cm and 37.5 cm but these figures are highly misleading because there are so many such areas where not even a single drop of rain is received for several years in continuation. Thus, the annual average rainfall is considered to be less than 12 cm. Most of Sahara receives less than 12 cm of mean annual rainfall. Cairo of UAR (3.0 cm), Lima of Peru (5 cm), William Creek of Australia (13.3 cm), Yuma of Arizona (USA, 8 cm), Toloth in S.W. Africa (5.6 cm) receive very low mean annual rainfall. Northern parts of Chile some times do not receive any drop of rains for 5 to 10 years in continuation. The equatorward relatively more humid areas, however, receive 50 to 75 cm as mean annual rainfall. Most of the rains is of convectional type due to local heating. Some times, occasional storms give heavy downpour within few hours causing flash floods. For example, 85 cm of rainfall was recorded within two days in Doorbazi of Rajasthan whereas its mean annual rainfall is only 12.5 cm. Such occasional catastrophic rainfall causes flash floods resulting into choking of storm drains, destruction of human settlements, silting and choking of canals and destruction and disruption of means of transport and communication. Such heavy rainfall is not useful at all as all of the rainwater are disposed off quickly through surface runoff and the remaining water is evaporated due to high temperature.

Skies are generally free from clouds and thus sun's rays reach the ground surface without being reflected throughout the year and hence the tropical desert climatic regions receive sufficient bright sunshine all the year round. The Sonoran Deserts of the USA and Mexico receive more than 75 per cent and 90 per cent sunshine during winter and summer seasons respectively. Most parts of Sahara Desert are characterized by 1/10 cloudiness in December and January and 1/30 cloudiness from June to October. It is, thus, apparent that the ground surface is more or less always baked in the sun. Some times, dark cumulo-nimbus clouds are formed, thunderstorms with cloud thunder and lightning are experienced but still there is no rain because raindrops are evaporated before they reach the ground surface. The average relative humidity ranges between 10 to 30 per cent.

Natural Vegetation

Hot desert type of climate is not conducive for vegetation growth because of acute scarcity of water. This is why most of the regions under this climate are either devoid of any vegetation such as Lybian and Arabian deserts or if there is any vegetation at all, that is very little, sparse and bushy in character. The vegetation of hot desert climate is of xerophytic type which has special characteristics to withstand harsh climate characterized by extreme aridity, high temperature and very high rate of evaporation. They have their own moisture conserving devices such as long roots, thick barks, waxy leaves, thorns and little leaves so that they may avoid evapotranspiration and consequent loss of moisture from them. Most of the vegetation are found in the form of bushes. Cactus, acacia, date palm, a few flowering plants etc. form the composition of natural vegetation of hot desert climate.

MEDITERRANEAN CLIMATE (CS CLIMATE)

Location

· The Mediterranean type of climates, climatically known as subtropical dry summer climates, is called Mediterranean type because most of the areas falling under this climate are situated around Mediterranean Sea. The Mediterranean climate or biome is also called as sclerophyl ecosystem or biome because of the development of special features and characteristics in the dominant trees and shrubs to adapt to the typical environmental conditions of this climate-dry summer and wet winter. Though the Mediterranean type of climate covers only 1.7 per cent area of the globe but this is most clearly defined climate and is easily differentiated from other climatic regions. The Mediterranean climate has three distinct characteristics–

(i) Wet winter and dry summer season,

(ii) Warm and hot summers, and mild winters, and

(iii) Abundant sunshine throughout the year (90 per cent in summer and 50 to 60 per cent in winter). This climate has developed between 30°-40° (some times upto 45°) latitudes in both the hemispheres in the western parts of the continents (Fig: 8.5). This climatic region includes

(A) European, Asiatic and African lands bordering the Mediterranean Sea *e.g.*

(1) European lands—Rhonesaone Valleys of France, southern Italy, Greece;

(2) Asiatic lands-western Turkey, Syria, Lebanon, western Israel;

(3) North coastal Africa—Morocco, northern Algeria and Tunisia and the area north of Bengasi in Lybia,

(B) Central and southern California in the USA,

(C) Central Chile in South America,

(D) The Capetown area of South Africa, and

(E) The coastal zones of southern and south-western Australia. The Mediterranean type of climate owes its origin to the seasonal shifting of wind and pressure belts due to northward (summer solstice) and southward migration (winter solstice) of the sun. Thus, these areas come under the domain of westerlies in winter season. Since the westerlies are moisture laden because they come from over the oceans and are associated with temperate cyclones, they give sufficient rains during winter season. On the other hand, they come under the influence of subtropical high pressure belt and associated anticyclones during summer season and hence there is no rainfall.

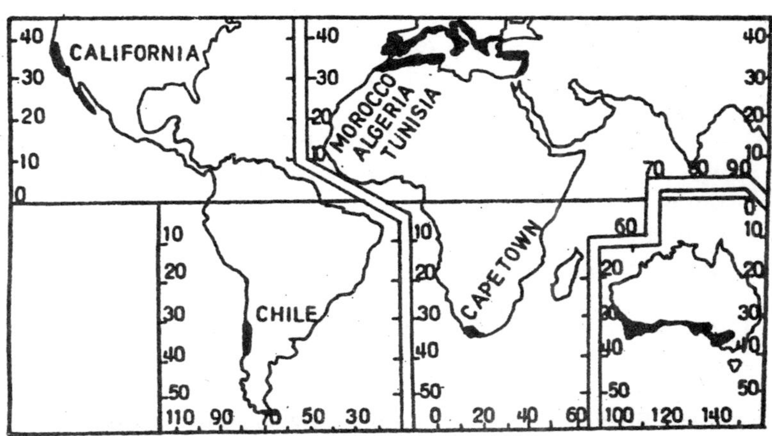

Fig. 8.5 : Spatial Distribution of Mediterranean Climate.

Temperature

The average temperature during cool winter season ranges between 5°C and 10°C whereas mean summer temperature varies from 20°C to 27°C and thus the mean annual range of temperature becomes 15°C to 17°C or even more. In fact, the Mediterranean climate is considered as a resort climate because of its pleasant and comfortable winter season. The Mediterranean climate whether having coastal or inland location generally records temperature above freezing point during winter season as the average temperature of the coldest month ranges between 4.4°C and 10°C. The mean January temperature is recorded as 7.7°C at Sacramento in California (USA), 6.1°C at Marseille (France), 12.8°C at Perth (Australia) and 6.6°C at Rome. Some times, the temperature becomes so low at nights that frost occurs which is very much injurious to field crops. The valleys and depressions have freezing to subfreezing temperatures in winter and hence valleys are avoided for sensitive crops like citrus fruits which are grown on hill slopes. Los Angeles (California, USA), Naples (Italy) and Sacramento (California, USA) have recorded lowest temperatures of –2.2°C, –1.1°C and –8.3°C respectively. It may be pointed out that occasional fall of temperature below freezing point is limited to a few minutes to a few hours only. Summer temperature rises above 26°C *e.g.* Red Bluff of Sacramento valley records 27.5°C whereas average temperatures during summer season are 26.6°C in European Mediterranean lands and 24°C in north-west Africa. It may be mentioned that high summer temperatures are never uncomfortable due to low relative humidity. Daily maximum temperature in summer goes beyond 26°C. For example, the Great Valley of California records daily maximum summer temperatures ranging between 30°C and 32°C. Red Bluff and Sacramento have recorded highest daily maximum temperatures of 45.5°C and 46°C respectively so far. The night temperature during summer season falls to 15.5°C, thus diurnal range of temperature becomes as high as 30°C. Daily and annual ranges increase from coastal areas to inland areas.

On the basis of temperature variations the Mediterranean climate is divided into 2 subtypes *e.g.* Csa and Csb. Csa climate characterized by hot summer, has inland location whereas Csb climate of mild summer is located along the coastal margins of the continents. It may be pointed out that temperatures of the Mediterranean climatic regions except those around Mediterranean Sea are largely affected by cool oceanic currents *e.g.* Califomian region by cool California current, Chilean region by cool Peru current. Cape Town region by Benguela currents etc.,

Air Procure and Winds

In fact, the Mediterranean climate is the outcome of the seasonal shifting of pressure and wind belts. All the pressure belts except polar high pressure belt move northward from their normal positions during northern summer (summer solstice when the sun becomes vertical over the tropic of Cancer on June 21) and thus subtropical high pressure system extends over the regions of Mediterranean climate (30°N to 40°N) and anticyclonic conditions dominate the weather conditions causing subsidence of air from above, horizontal divergence of surface winds, stability of air resulting into dry conditions. Sirocco local hot winds blow from Sahara northward picking red sands which are brought to Italy, Spain, southern France and Greece. Summer winds are generally hot and dry. These areas come under the domain of westerlies during winter season when the sun moves southward (northern winter). The westerlies are associated with temperate cyclones originating in the middle latitudes. Since the westerlies are associated with temperate cyclones and come from over sea surfaces and hence being moist give sufficient rainfall in the coastal areas. Some local winds like Bora. Mistral (cold winds) etc. affect the local weather conditions of the European Mediterranean lands during winters.

Precipitation

The mean annual rainfall ranges between 37 cm and 65 cm, the most portion of which (more than 75 per cent) is received during winter season mainly between December and March in the northern Hemisphere and between May to September in the southern hemisphere. The winter rainfall is received through cyclonic storms associated with moist westerlies. The summer season is almost dry. Because of moderate to scanty rainfall the

Mediterranean climate is called as subhumid climate. The orographic rainfall is common in those coastal areas (*e.g.* California) which are backed by mountain ranges parallel to the coasts. Though winter season is quite wet but the sky is seldom overcast for longer duration in continuation and thus abundant sunshine is available even during wet winter season. For example. Red Bluff (California, USA) has only 11, 12, 10 and 10 rainy days in December, January, February and March respectively. January is the rainiest month in San Bernardino and Los Angeles but there are only 7 rainy days. January being the coldest month of Red Bluff receives 11.5 cm of rainfall while July being the coldest month of Perth (Australia) receives 16 cm of rainfall. All these indicate the intensity of rainfall during winter months in the Mediterranean climatic regions. Summer season is characterized by scanty rainfall, almost dry weather, clear skies, and bring sunshine. Besides temporal variation, there is also spatial variation of rainfall. Generally, the amount of rainfall increases from south (24 cm at San Deigo, 29cm at Los Angeles) to north (58 cm at Sansfransisco) in California.

The seasonal regime of rainfall causes fluctuations in the soil-water and soil-moisture regimes during winter and summer seasons. The amount of soil-water increases during winter season because of winter and spring rainfall which is responsible for maximum growth in vegetation but dry .summer season causes deficiency in the soil-water content because of loss of water and moisture due to increased evaporation and evapotranspiration because of substantial increase in temperature and of course due to general lack of rainfall during summer season.

Natural Vegetation

Though the Mediterranean regions are widely scattered over different continents, there is more or less broad generalization in the overall structure and composition of vegetation community of all the regions of Mediterranean biome. The structure of Mediterranean vegetations is such that they can withstand the aridity of summer season. Consequently, the leaves have developed sclerophyllous characteristics in that they are stiff and hard and the stems have thick barks. The

Mediterranean vegetation community consists of a variety of sclerophyll plant formation classes which range from Mediterranean mixed evergreen forests (in the coastal lands immediately bordering the areas and the oceans) to woodland, dwarf forest and scrubs. The vegetation community is dominated by trees and shrubs. The shrubs are differently named in various parts of the Mediterranean biome on the basis of local names *e.g.* maquis or garigue in southern Europe, chaparral in California, fymbos or tymbosch in South Africa, and malice scrub in Australia.

The xeromorphic structure such as thickened suiticles, grandular hairs, sunken stomata etc. enables the plants to withstand dry conditions. The sclerophyllous structure of plant leaves enables them to regulate the gaseous exchange according to the availability or scarcity of water during different seasons of the year. The plants have also developed special types of root systems in accordance with the regional environmental conditions mainly the availability of moisture.

The European Mediterranean regions are characterized by multi-layered structural pattern of vegetation community consisting of (i) topmost layer of evergreen and deciduous oak trees. (ii) middle layer of shrubs locally called as maquis or garigue, and (iii) the ground layer of numerous herbaceous plants. The Californian Mediterranean lands are characterized by the (i) topmost layer of oak trees, (ii) the middle layer of chaparrals, equivalent to European maquis, and (iii) the ground layer of herbaceous plants and grasses. The South African Mediterranean Biome is characterized by attractive flowering plants of numerous varieties *e.g.* Erica, Ereesia, Lobellia, Kniphofia species etc. The shrubs are locally called as fymbos. The Australian Mediterranean Biome is characterized by numerous species of eucalyptus.

Fire, both natural and man-included, is normally an annual occurrence in the Mediterranean climatic regions. Burning, mass clearance of natural vegetation for agricultural and commercial purposes, overgrazing of grasslands etc. have led to accelerated rate of soil erosion, increase in the silt load of major rivers and transformation of original natural vegetation.

CHINA TYPE OF CLIMATE (CA)

(Subtropical Humid Climate)

China type of climate, climatologically known as subtropical humid climate, is characterized by hot summer, mild to cold winters, spatial variations in temperature, humility and precipitation and is located between 20°-400 latitudes in both the hemispheres along the eastern parts of the continents (Fig. 8.6). This climatic region is flanked by inland dry regions in the west (except in Po and Danube basins), by monsoon climatic areas in the south, by humid continental climate in the north, and by oceans in the east. The subtropical humid climate (Ca) is found in the following regions-south-east and south China (south of Yellow river); Po Basin; Danube Basin; south-eastern USA; south-eastern Brazil, Paraguay, Uruguay and north-eastern Argentina; south-eastern Africa; and east coast of Australia.

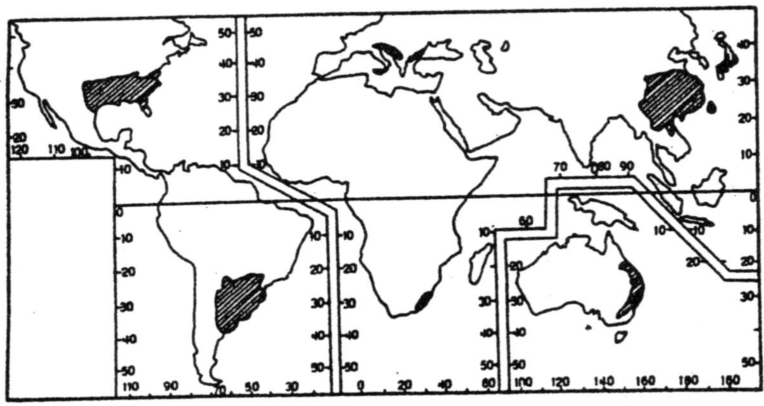

Fig. 8.6 : Distribution of China type of climate.

It is apparent that Mediterranean type and China type of climates are found in the same latitudinal locations but are differentiated on the following counts-(i) Mediterranean climate is found in the western parts of the continents while China type of climate is located along the eastern coastal areas of the continents, (ii) Mediterranean climate is characterized by wet winter and dry summer whereas in China type of climate summer season receives maximum rainfall though rainfall is received throughout the year. (iii) Mean annual rainfall is higher in China type of climate than Mediterranean climate. China type of climate is also called as sub monsoon climate because of its near similarity with monsoon climate.

Temperature

The coastal parts of China type of climate are frequented by warm oceanic currents (*e.g.* warm Gulf Stream off the east coast of the USA, warm Brazil current along the east coast of South America, warm east Australia current off the east coast of Australia, Kuro Shio current off the coast of China, and Mozambique current along the south-east coast of Africa) and thus these warm currents affect the temperature of coastal areas. The mean summer temperature ranges between 24°C and 26.6°C. The month of July records mean temperature of 28°C at Charlston (South Carolina, USA) and Montgomery (Alabama, USA), 27.2°C at Shanghai whereas January, the wannest month in the southern hemisphere, records mean temperature of 25°C at Brisbane (Australia) and Durban (Africa), and 23.3°C at Buenos Aires (South America). Daily maximum temperature crosses 37.7°C (100°F). For example, maximum daily temperature at Montgomery is recorded as 41°C in July and 41.6°C in August. Similarly, Savannah of Georgia (USA) has recorded 41°C as maximum daytime temperature in July. It may be pointed out that high summer temperature is also associated with high relative humidity and hence the weather becomes uncomfortable.

The loss of the heat during nights through outgoing longwave radiation is retarded because of cloudiness and hence temperature does not fall appreciably at nights resulting into low diurnal range of temperature. For example, the mean maximum and mean minimum temperatures of Montgomery (Alabama, USA) are 32.8°C and 22.2°C respectively and thus diurnal range of temperature becomes 11.6°C.

Generally, winters are mild as mean temperature ranges between 6.6°C and 10°C. Mean winter temperature is recorded as 9.4°C at Montgomery (Alabama, USA), 3.3°C at Shanghai, 10°C at Buenos Aires, 11.7°C at New Orleans (USA), 6.1°C at Nagasaki, and 11°C at Sydney. Mean annual range of temperature in China type of climate is not large but is marked by wide range of spatial variations. For example, mean annual range of temperature is 12.8°C at Buenos Aires and Sydney, 10.5°C at Montgomery and 24°C at Shanghai. In fact, annual range of temperature is controlled by the extent of land areas and the strength of winter monsoon winds. The more extensive the land areas and stronger the winter monsoon, the larger is the annual range of temperature.

The arrival of cold polar winds during winter season causes temperature to fall below freezing point. For example, in the absence of any mountain barrier from west to east in North America, the extremely cold continental polar airmass invades the whole of the plains of the USA and reaches the Gulf Coastal plains of the country and thus the temperature falls below freezing point (about –12°C) in the Gulf Coastal state of the southern USA during winter season.

Air Pressure and Winds

The continental and oceanic areas are characterized by low and high pressure systems during summer season, with the result monsoons are developed. Thus, tropical maritime airmass (mT) develops over land areas falling in Ca climate mainly over China and southeastern USA. This air mass is south-easterly in the northern hemisphere while it is north-easterly in the southern hemisphere (during summer season).

High pressure systems develop over south-western and northern central Pacific Ocean and near Azores in the Atlantic Ocean. These tropical maritime airmasses are also associated with tropical cyclones which yield sufficient rainfall. These tropical cyclones are called typhoons in China, hurricanes in the USA and southern busters in Australia.

The pressure system is reversed during winter season as high pressure develops over continental areas while oceanic areas are characterized by low pressure systems. Thus, the winds become offshore. This is the reason that the coastal areas of the eastern and south-eastern USA are not much benefitted from the warm Gulf Stream because the winds are offshore.

Precipitation

Though average annual precipitation in subtropical humid climate (Ca) ranges between 75 cm and 150cm and some times it becomes as much high as 250 cm in some favoured locations but there is wide range of variation in the seasonal and spatial distribution of annual rainfall. Generally, rainfall decreases from coastal areas to the inland locations. On the basis of seasonal distribution of rainfall this climate is divided into two subtypes *e.g.* (i) Caf climate having rainfall throughout the year, and (ii) Cw climate having dry season during winter.

On an average, the summer rainfall in China type of climate is definitely abundant and exceeds evapotranspiration. The summer rainfall is received through cumulus clouds resulting from convective currents caused by intense local heating of the ground surface. Besides, tropical cyclones (*e.g.* typhoons in China and hurricanes and tornadoes in the USA) also yield heavy downpour with cloud thunder and lightning. The Gulf coastal states of the USA are generally frequented by large number of thunderstorms. For example, Florida experiences at least 60-thunderstorms each year.

These thunderstorms are associated with unstable tropical maritime air mass. The summer rainfall in China and Japan is divided into three types of regime viz. (i) maximum rainfall period of early summer season, (ii) maximum rainfall period of late summer season, and (iii) minimum rainfall period of middle summer season. Some times, the rainfall occurring from hurricanes in the south-east USA and from typhoons in China is so heavy that catastrophic floods are caused.

Hurricanes very often strike the southern and south-eastern coasts of the USA. The Gulf coasts of Louisiana, Texas, Alabama and Florida are worst affected areas. The Galveston, Texas (USA) disaster of September 8,1900 tells the story of devastation caused by hurricane in Gulf coastal states of the

USA. The terrible hurricane generated a strong storm surge (tidal wave) which raced inland and killed 6000 people mostly through drowning caused by inundation under 3 to 4.5m deep water.

Winter rainfall is generally received through winter cyclones which are associated with the westerlies. Though the duration of individual rainstorm during winter season is much longer but total rainfall is comparatively less. It may be pointed out that inspite of less rainfall during winter season there is more cloudiness and larger number of rainy days than during rainier summer season. For example, 5 cm of January rainfall of Shanghai is received within 12 days whereas 15 cm of August rainfall is received in only 11 days. Some times, there is occasional frost and snowfall during winter season.

Natural Vegetation

The China type of climate characterized by abundant rainfall, high temperature and long growing season of 7 to 12 months favours luxuriant growth of natural vegetation. Dense forests of evergreen nature are found in more humid areas but areas of moderate rainfall are characterized by deciduous sparse forests and grasslands. Normally, mixed forests of coniferous trees and broad leaf trees are found. The broad-leaved forests are both evergreen and deciduous depending on the spatial variation of distribution of annual rainfall.

STEPPE CLIMATE (BSK)

Location

The middle latitude steppe climate (BSk) spread over temperate grasslands is located in the interiors of the continents which come in the westerly wind belt but because of their more interior locations they do not get sufficient rainfall and hence the grasslands are practically treeless. The temperate grassland steppes of the southern hemisphere are located along the south-eastern margins of the continents (Fig 8.7) and therefore have more moderate climate than their counterparts of the northern hemisphere because of more marine influences as they are closer to marine environments.

The temperate grasslands of Eurasia, known as steppes, are most extensive as they extend for a distance of more than 3200 km from the shores of the Black Sea across the Great Russian Plain to the foothills of the Altai Moutanins. Their continuity is broken at few places by the highlands. There are also some isolated patches of steppes *e.g.* in Hungry (known as *Pustaz*) and in the plains of Manchuria (Manchurain Grassland). The temperate grasslands in North America (extending in Canada and USA both) are locally known as *prairies* which extend from the foothills of the Rockies in the west to the temperate deciduous forest biome in the east.

The temperate grasslands of the southern hemisphere include the *pampas* of Argentina and Uruguay of South America, *bush veld* and *high veld* of South Africa, and downs of the Murray-Darling Basins of southeastern Australia and *Canterbury grasslands* of New Zealand (Fig. 8.7).

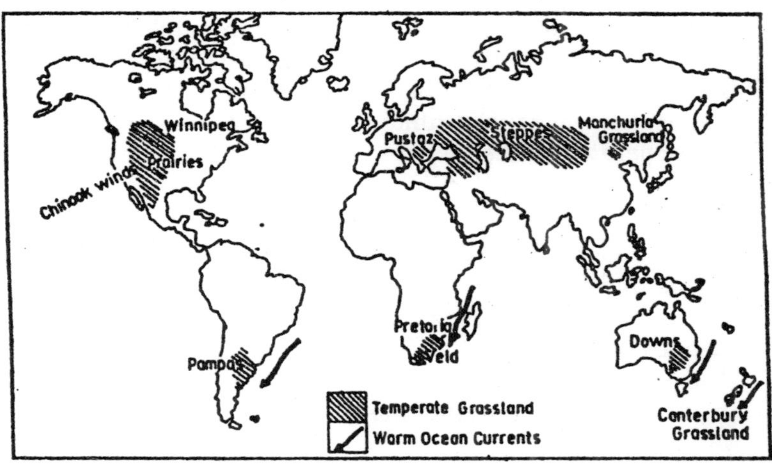

Fig. 8.7 : Spatial distribution of temperate steppe or grassland climate.

Temperature

The temperate steppes are characterized by continental climate wherein extremes of summer and winter temperatures are well marked but the temperate grasslands of the southern hemisphere are marked by more moderate climate. Summers are warm with over 20°C temperature in July (Winnipeg, Canada) and over 22°C in January (Petoria, South Africa,

January is summer month in the southern hemisphere). Winter season becomes very cold in the northern hemisphere because of enormous distances of temperate grasslands from the nearest sea. Winnipeg records –20°C in January. The average temperatures during winter season for Tashkent (Uzbeck Republic) and Semipalatinsk (Kazakh Republic) are 0°C and –12°C respectively. The steppe climate of the southern hemisphere is never severe rather it is moderate because of nearness to the sea. The average winter temperature ranges between 1°C and 12°C in the southern hemisphere.

The steppe climate is characterized by high annual range of temperature. For example, Winnipeg (Canada) records mean annual range of 40°C. Laramine (Wyoming state in USA) records annual range of 23.4°C (–6.7°C in "January and 16.7°C in July). Due to marine influences the mean annual range of temperature in the southern hemisphere is much lower than the northern hemisphere as it is around 10°C to 12°C only. Diurnal range of temperature is also very high in the temperate steppe climate.

Precipitation

The mean annual precipitation ranges between 25 cm to 75 cm in different locations of the temperate grassland steppe areas. The winter precipitation in the northern hemisphere is usually received in the form of snowfall and most parts of Eurasian steppes are snow-covered for several months during northern winters. Most of the annual rainfall is received during summer season.

Regional Characteristics

The Eurasian steppe climate covers the largest area in the former Soviet Union (now the Commonwealth of Independent States-CIS and other Republics of former USSR) wherein it extends from eastern Europe to western Siberia and between temperate coniferous forests in the north and arid regions in the south-west. The Eurasian steppes are divided into (i) forest steppe, and (ii) grass steppe wherein the former receives mean annual precipitation of 50 cm to 60 cm whereas the latter receives 40 cm to 50cm as mean annual precipitation. The following sequences of vegetation communities are found from north to south–

(1) Forest steppe (consisting of oak, elms, limes, mapple, a few arboreal vegetation of Siberian Steppe such as birch with mixture of aspen and willow),

(2) Meadow steppe (consisting of the species of turf grasses such as stipa and Fescue and numerous herbaceous flowering plants such as numerous herbaceous flower plants such as Trifolium and several types of daisy,

(3) Grass steppe (consisting of grasses mainly tussock-forming species of Stipa, a few flowering xerophytic shruby species of Artemisia,

(4) Semi-arid xeromorphic steppe dominated by xerophytic grasses such as Fescue and feather grass species, (mean annual precipitation ranging between 30 cm to 35 cm).

The North American Prairie has developed in the USA and Canada between the foothills of the Rockies in the west and temperate deciduous forest biome in the east. On the basis of decreasing trend of mean annual precipitation from east (105 cm) to west (40 cm) the North American Prairies are divided into 3 subregions *e.g.* (1) tall grass prairie (most dominant species of tallest grass are Bluestem and Switch Grass which attain the height of 1:5m to 2m, few patches of oak and hickory trees are also found), (2) mixed prairie (most extensive cover in the Great Plains of the USA, mixture of medium grasses, 0.6 to 1.5m in height, and short grasses such as little bluestem, needle grass-Stipa spartea, June grass and short and bunch grasses such as buffalo grass and blue gramma, and (3) short grass prairie (developed over western part of the Great Plains and dominated by short grasses of 60 cm height).

The South American Pampas are developed over 12 per cent area of Argentina and are more humid than Eurasian steppes. The mean annual precipitation decreases from east (coastal land, 90 cm) to west (45 cm). Thus, the Pampas are divided into two sub-types *e.g.*, (i) Humid Pampa and (ii) Subhumid Pampa. The humid Pampa, developed in the eastern part of Argentina, is characterized by tall grasses whereas increasing aridity westward results in the growth of short grasses in the western sub-humid Pampa.

The African veld has developed on the high plateau land of varying heights (1500m to 2000 m) in the south-eastern part of South Africa. The African Velds include the temperate grasslands of southern Transwall and Orange Free State of South Africa and some parts of Lisotho. Plant growth is not possible because of uncertainty of rainfall, increasing aridity, severity of frosts during nights and high daily range of temperature during winter season and thus true climax grasslands of African Velds have developed.

The Australian downs have developed in the south-eastern parts of Australia and in the northern part of Tasmania. The region is characterized by (i) relatively warmer winter season than the temperate grasslands of the northern hemisphere, and (ii) mixture of grasses with eucalyptus trees. The grasslands gradually change from south (Australian coast) to north (interior land) in accordance with the decreasing trend of mean annual precipitation from south (152 cm) to north (63.5 cm). The region is further divided into distinct 3 subregions *e.g.* (1) temperate tall grasslands; (2) temperate short grasslands, and (3) xerophytic grasslands.

WEST EUROPEAN TYPE OF CLIMATE (CB)

Location

West European type of climate (Cb) also known as *marine west coast climate* is located between 40° and 65° latitudes in both the hemispheres along the western coasts of the continents. This climatic region is surrounded by Mediterranean climate in the south, continental dry climate in the east and semi-arctic climate in the north. The inland extension of this climate is controlled by topographic features. For example, wherever the coast is paralleled by mountain ranges, this climate is found in a very narrow coastal belt *e.g.* marine west coast climate is confined to the coastal strips along the western coasts of North and South Americas because of Rockies and Andes. On the other hand, wherever relief barrier does not exist, marine influences reach far inland *e.g.* north-western Europe. Thus, the west European type of climate has developed over north-western Europe (including Great Britain, western Norway, Denmark, northwest Germany, Netherlands, Belgium, Luxemberg, and north-western France), British Columbia of Canada, Washington and Oregon states of the USA, south-west coast of Chile (S. America), south-east coast of Australia, and Tasmania and New Zealand.

Temperature

The temperatures, in the west European climate are affected by marine influences, warm ocean currents and prevailing winds and air masses. In fact, the moderating effects of sea bring down the difference between summer and winter seasons considerably. This climate is characterized by cool summer and mild winters. Average temperature during summer season ranges between 15°C and 21°C. Thus, the summer months are characterized by negative thermal anomaly *i.e.,* the coastal regions in the marine west coast climate record relatively lower temperature during summer season than the average temperature for their latitudes. There is very negligible variation in the spatial distribution of temperature during summer season as it is indicated by mean July temperature of the following stations = 17°C at Seattle (USA), 14.4°C at Bergen (Norway), 15.6°C at Dublin (Ireland), and 19°C at Paris (Prance). The daily minimum temperatures in July at Seattle and Bellingham (Wales) are 12.8°C and 10.6°C respectively while daily maxima at these stations are 22.8°C and 21.7°C respectively. Thus, diurnal range of temperature in July for Seattle and Bellingham becomes 10.0°C and 11.1°C. Some times, daytime summer temperature exceptionally rises to 32°C-38°C.

Winters are exceptionally milder for their latitudes due to proximity of warm ocean currents and thus the coastal locations of western Europe are characterized by positive thermal anomaly *i.e.,* they record higher temperature than the average temperature of their respective latitudes due to the influence of warm North Atlantic Drift (extension of warm Gulf Steam). The positive thermal anomaly of 11°C to 17°C is a common feature. The winter temperature decreases rapidly from the coasts towards the interior parts in Europe due to decreasing marine influence inland. This is why the January isotherms instead of following latitudes become parallel to the coasts. The mean January temperature in coastal areas of N.W. Europe ranges between 4°C and 10°C but it becomes –18°C to –40°C in the interior continental locations of Eurasia. The night temperature generally falls below freezing point and hence ground frost is of very common occurrence. Cold waves are generated due to arrival of cold continental polar air masses.

The marine west coast climate comes under the domain of westerlies which are regular features throughout the year. Since these winds come from over the oceans and hence they are moist and give precipitation. These westerlies are also associated with temperate cyclones which are the main sources of precipitation. The poleward margins are dominated by subpolar low pressure belt of dynamic origin where unstable polar front is formed due to convergence of two contrasting air masses *e.g.* warm and moist westerlies and cold polar air mass. This polar front thus causes the development of temperate cyclones which move in easterly direction under the influence of westerlies.

Precipitation

Marine coast climate or West European type of climate is basically humid climate and is characterized by abundant and uniformly distributed precipitation throughout the year cut winter maximum is the characteristic feature of coastal locations while interior locations record summer maximum. Inspite of abundant precipitation all the year round there is much spatial variation in its amount. Generally, precipitation decreases from the coasts towards interior locations and from north to south along the coast. The regional distribution of precipitation is highly controlled by topographic factor. The areas of low reliefs receive relatively low precipitation. For example, the north-western European lowland in the absence of any effective relief barrier receives mean annual precipitation ranging between 50 cm and 75 cm. On the other hand, the western coastal areas of North America and of Chile in South America falling under marine west coast climate receive high mean annual precipitation ranging between 250 cm and 375 cm because of the presence of the coast range mountains in North America (parallel to the coast) and the Andes in South America (parallel to the coast). The leeward slopes of these mountains become dry because of very low precipitation as they fall in rainshadow region.

Though the precipitation is uniformly distributed throughout the year but winter season receives more than the summer season, but there is no dry month. These conditions are confined only to the coastal location because interior locations receive more precipitation in summers than in winters. The precipitation in low land areas (plains) is cyclonic in nature and is usually received in the form of drizzles and continues for fairly long time. Sky remains overcast for several days in continuation. The winter cyclonic precipitation is very widespread. The summer precipitation is of short duration but is stormy and heavy. The 6 cm July precipitation of London is received in 13 days while 5.3cm January precipitation comes in 15 days. The thunderstorms are very few in number.

It is interesting to note that though the mean annual rainfall is moderate but it is received in large number of rainy days. For example, the mean annual precipitation of 56.5 cm of Paris is received in 188 rainy days while London gets mean annual precipitation of 71.3 cm in 164 days. The percentage of cloudiness is also much higher in this climate *e.g.* the Pacific coastal areas of N. America record average annual cloudiness between 60-70 per cent while it is 70 per cent in the western Europe. Winter months are also characterized by snowfall but the number of days receiving snowfall is less than in other climates located within the same latitudes. The snow days in London, Paris and Seattle (representing low land location) are 13, 14 and 10 respectively. The frequency of snow-days and intensity of snowfall both increases poleward and towards interior locations.

Natural Vegetation

The abundant precipitation throughout the year has given birth to dense forests of three types *e.g.* (i) broad-leaf deciduous forest (oak, birch, walnut, maple, elm, chestnut etc.), (ii) needle-leaf (coniferous) forest (pine, fir, spruce etc.), and (iii) mixed forest but cleared in British Isles and European countries due to urban and agricultural development. Dense forests are now found only on mountains and highlands. Douglas fir, redwood, hemlock, spruce, cedar etc. soft wood forests of much commercial use are found in the states of Washington, and Oregon of the USA and British Columbia of Canada.

BOREAL OR SUBARCTIC OR TAIGA TYPE OF CLIMATE

Location

The boreal or sub-arctic climate representing the boreal forest biome or temperate coniferous forest biome and the most extreme type of microthermal climate is called taiga type or Siberian type of climate and includes the areas of subarctic regions of North America (extending from Alaska of the USA across Canada to Hudson Bay in the east) and Eurasia (from the Scandinavian Peninsula across the Russian Siberia to the Bering Sea) (Fig. 8.8). Besides, there are small patches of this climate at higher altitudes in Germany, Poland, Switzerland, Austria and other parts of Europe and on the high Rocky Mountains of North America. In fact, the taiga climate is located between the tundra climate in the north and the temperate grassland biome (climate) (Eursian steppes and North American Prairies) in the south. The taiga climate is conspicuous by its total absence in the southern hemisphere because of narrowing trend of continents towards the south pole. The vicinal location of taiga climate extends from 50°-55°N to 650-70()N latitudes.

Temperature

The taiga type of climate is characterized by extreme continental climate marked by bitterly cold winter of long duration and cool short summer season of brief period extending over one to three months. Spring and autumn are merely brief transitional periods between summer and cold seasons. The 10°C isotherm of the warmest month forms the northern boundary of this climatic region. The winter season extending over 8 months always records temperatures below freezing point as is apparent from the average January temperatures of the following inland locations = –26°C at Eagle (Canada), 30.6°C at Dawson (Canada), –24°C at Okhotsk (Russia), 43.3°C at Yakutsk (Russia), –50.6°C at Verkhoyansk (Russia) etc. The Siberian taiga climate records the lowest minimum temperature *e.g.* Verkhoyansk –68°C and Oimekon –66.8°C (Russia, lowest temperature ever recorded) while the lowest minimum temperature recorded so far in North American sub-arctic climate at Snag (Yukon, Alaska, USA) is –62.8°C.

In comparison to severe cold winter months temperature during brief summer season increases rapidly. July being the warmest month has an average temperature of 16°C. It may be pointed out that several interior locations record temperature below freezing point even in the month of July. The growing season is between 50 to 70 days only because soil water is frozen for 5 to 7 months of winter season in continuation. The annual ranges of temperature are very large and greatly vary from place to place. For example, the temperature of the coldest and warmest months of Moscow are –12°C and 20°C respectively and thus the annual range of temperature becomes 32°C. Verkhoyansk records the annual range of temperature of more than 64°C.

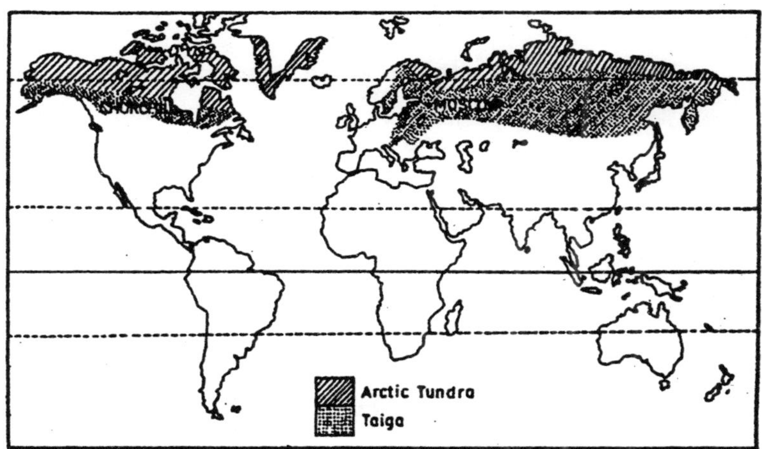

Fig. 8.8: Location of taiga and tundra types of climate,

Precipitation

The subarctic or taiga type of climate is characterized by low mean annual precipitation because (i) extremely low temperature for longer period of the year does not favour evaporation and thus there is very low amount of absolute humidity in the air, (ii) the regions falling in taiga type of climate are located in the leeward sides of the continents and in the interior regions and thus they are away from the marine influences, and (iii) the regions are characterized by polar high pressure, anticylonic conditions as evidenced by subsidence of air from above and divergence of surface winds. The mean annual precipitation ranging between 37 cm and 60 cm is received mostly in the form of fine, dry snow which accumulates throughout the winter and is released as surface water due to thawing because of increase in temperature during brief summer season. It may be pointed out that the precipitation is more or less uniformly distributed throughout the year whether in liquid form as rainfall (during summer) or snowfall during winters.

The following are the main characteristic features of subarctic taiga type of climate–

(i) Bitterly cold long winter season (temperature below zero degree centigrade at least for 6 months).

(ii) Heavy snowfall during winter season.

(iii) Formation of permafrost ground (permanently frozen ground) because of freezing of ground moisture due to subfreezing temperature.

(iv) Cool summer of short duration having precipitation in liquid form (rainfall) and melting of snow cover.

(v) Growing period of vegetation ranging from 50 days a year along the northern boundary to 100 days along the southern boundary.

(vi) High range of variability in the spatial distribution of annual precipitation

(vii) Extreme annual range of temperature (ranging between 25°C during summer months and 40°C during winter months).

Natural Vegetation

The coniferous trees are the most dominant member of taiga climate or the Boreal Forest Biome. These trees form dense cover of forests which are the richest sources of softwood in the world. There are four major genera of evergreen coniferous trees *e.g.*, (i) Pine (white pine, red pine, Scots pine, jack pine, lodgepole pine etc.). (ii) Fir (doughlas fir, balsam fir etc.), (iii) Spruce (Picea), and (iv) Larch (Larix). Besides, a few species of temperate deciduous hardwood trees have also developed in this climate mainly in those areas which have been cleared by man through felling of original temperate coniferous trees. Thus, the temperate deciduous trees represent the 'secondary succession of vegetation' which includes alder, birch and poplar.

GENERAL FEATURES OF CLIMATES

Climate directly affects the daily life of every person on the earth's surface, and forms an important feature of the environment. There is admittedly much controversy concerning the relationship between climate and racial characteristics, but no one can deny that some relationship does exist.

Climate helps to determine food, clothing, housing and general mode of life; its effects encourage or discourage the diseases and pests to which man is exposed in various parts of the world. The trade routes which opened up the world in sailing-ship days were controlled to a large extent by air currents, which in the air age are still important. Moreover, climate affects other features of the environment. Climatic elements are largely responsible for the detailed sculpture of the face of the earth—weathering, the work of running water and of glaciation, wind and water action in desert lands, even the storm-waves which pound the coasts, are all the results of climate. The mantles of soil and vegetation, both natural and cultivated, owe much of their character to their climatic environment.

Climatic Changes

The various climates of the world today differ considerably from those in the past. Many parts seem to have experienced both 'warm periods' and 'cold periods' (commonly called 'Ice Ages'), and also 'arid periods' and 'pluviose periods', as has been determined from such geological evidence as the nature of the rocks themselves and the character of the remains, plant and animal, found therein. The general distinction between Pole and Equator seems to have obtained for a very long time, but at times polar influences extended equatorward, at others the reverse took place. A mid-latitude location, such as the British Isles, has experienced a succession of changes—a tropical climate in the Carboniferous period, a desert climate in the Triassic, a cool temperate climate in the Cretaceous, and an arctic climate in the Quaternary, succeeded by minor fluctuations in postglacial times.

These climatic changes are not limited to long-distant geological time. A fascinating line of research is the tracing of climatic changes within historical times by piecing together evidence from a variety of sources, many items scanty in themselves but often supplementing and corroborating some other information. There is much evidence of ruined settlements in what is now desert, of ancient irrigation works, of shrinking glaciers, of the width of the annual growth rings in trees (dendrochronology), and of both legends and eyewitness records of floods, droughts and great frosts. One example of such evidence is the discovery of coffins in Greenland, exhumed from what is now permanently frozen terrain, yet with remains of tree roots penetrating the coffins. When all the available evidence is analysed and put together with care, a climatic time-chart for the last few thousand years can be constructed.

CLIMATOLOGY AND METEOROLOGY

Weather and Climate

The term weather is used to describe the condition of the atmosphere at a certain place at some specific time; it is essentially a day-to-day, or even an hour-to-hour, phenomenon. Climate comprises a description of the condition of the atmosphere over a considerable area for a long time. Thirty-five years was usually considered to be the minimum period of time which must elapse before an adequate summary of a particular climate can be made. In some parts of the world this 'climate' is harder to describe in average terms than others; thus it is sometimes said that the British Isles experience weather rather than climate.

Climatology The science of climatology falls largely within the province of the geographer, since it involves an analysis of the spatial distribution of atmospheric phenomena. A climate has been called a 'geographical entity', and much of the work of the climatologist lies in defining and describing different climates in their geographical settings. The terms 'climate' and 'climatic type' are commonly used synonymously; the description of the major climatic types forms the basis of Chapter 18. Climatology is in part descriptive, presenting an actual survey of pressure, temperature, rainfall and the other elements, but it must to some extent be explanatory as well. Here, then, the geographer comes into contact with meteorology, and he has realized in recent years that he must draw more and more from this science to assist his understanding of the climatic elements, hence the terms 'synoptic' and 'dynamic' climatology, whereby the subject is studied from the view-point of its relationship to the pattern of atmospheric energy transfer and exchange, of heat and water balances, and of the general circulation of the atmosphere.

Meteorology

Meteorology is the scientific study of the physical processes constantly at work in the atmosphere. Modern meteorology is to a large extent the field of the mathematical physicist, for a training in thermodynamics and hydrodynamics is essential if atmospheric processes are to be fully understood. One function of meteorology is its application to the forecasting of the future trend of the weather, a development of vital importance in this age of aviation. The basis of conventional forecasting is the construction of synoptic charts from simultaneous observations of atmospheric phenomena at a number of stations. Modern developments include mathematical forecasting, using a computer to solve sets of 'forecasting equations'; the recognition of periodicities and singularities', and in long-range forecasting the use of analogues, that is, the deduction of a series of future repeating patterns of weather situations by analysing and comparing sequences of past years.

More important from the point of view of the climatologist is the contribution of scientific method and explanation which meteorology makes to his descriptive distributions. The analytical study of individual masses of air with distinctive features of temperature and humidity, and of the conditions of the upper air, have both yielded much information which the geographer can use to assist his climatological descriptions.

The data The data of climatology are provided by stations which have recorded temperature, pressure, wind, rainfall and the other elements over a period of years. In western Europe and eastern North America there is a reasonably close network of such stations, elsewhere they are much more sporadic; there are about 2500 stations in the northern hemisphere. There are many parts of the land-masses (and much of the oceans) for which there are only short-term records or none at all; much of the work of the climatoologist, in describing climatic types, must be in the form of cautious deductions from general principles and short-term observations. In recent years, knowledge of oceanic meteorology has been substantially increased by the systematic compilation of records by permanently stationed 'weather-ships', maintained by the U.S.A., Great Britain, Norway and other European countries in the eastern Atlantic, and by flights of specially equipped aircraft, such as the Research Flight for the Meteorological Office. Numerous weather-satellites in orbit now transmit continuous information.

In January, 1967 a high-speed weather communication link was opened between North America and western Europe, transmitting information from satellites in orbit. This link forms part of the World Weather Watch established by the World Meteorological Organization, with its headquarters in Geneva, which began operation in 1968. The heart of the system consists of three centres at Suitland (Maryland, U.S.A.), Moscow and Melbourne, the clearing houses for vast bodies of statistical information concerning world weather patterns from sea-level to about 30 km (20 miles) in altitude. The whole essence of this elaborate organization is based on the development of three techniques: high-speed long distance communication links, electronic computers and orbiting satellites providing not only photographs of changing cloud patterns but also vast quantities of data for computer processing.

'LOCAL CLIMATE' AND MICROCLIMATOLOGY

In recent years there has been considerable development in the detailed study of 'local climates'. This involves the careful examination of slight, yet significant, contrasts in climate that may result from small differences of slope and of aspect, from the colour and texture of the soils, from the proximity of water surfaces, from the nature of the vegetation cover, and from the effects of buildings.

Its application involves the appreciation of the significance of these small climatic variations for plants, animals, insects and human beings. The vital question of frost incidence, for example, has occupied the attention of climatologists and horticulturists alike. Various lines of research include the routes taken by the downward creep of cold air, the position of 'frost pockets', and the creation of 'frost-breaks' in the form of thick hedges. There has been much research into urban climates—the study of the interference of buildings with air movement, the questions of atmospheric pollution and of fog occurrence, and the degree of warming of the mass of air resting in and above a city (*e.g.*, London's 'heat-island'). The siting of new towns, villages and airports calls for detailed preliminary work in this field. This mainly involves the recording of the temperature of the air layer, a few metres thick, resting on the ground, and includes records of soil temperature at certain depths.

A distinction is made between the study of a 'local climate', that is, the climate of a small area, and of a 'microclimate', that is, the detailed and very small-scale study of meteorological conditions within natural and cultivated vegetation, principally from an ecological point of view. A further term now employed, logically enough, is micrometeorology, which implies the detailed scientific study of the lowest layer of the atmosphere.

CLIMATIC FACTORS AND ELEMENTS

The term *climatic element* is a convenient label for each of the constituents which make up the sum total of climate—temperature, air pressure, wind, humidity and precipitation. These elements result from the interplay of a number of *factors or controls*, that is, determining causes. In the description of each element which follows, the applicable determining factors are also considered. Several factors appear over and over again.

Latitude is an important factor, since it determines both the length of day throughout the year and the intensity and possible duration of sunlight received. Altitude has very definite results on temperature, pressure and precipitation, and mountain ranges sometimes form clearly defined climatic barriers and therefore boundaries of climatic types. The distribution of land and sea is a third factor; 'continental' and 'marine' influences can be very important. Ocean currents exercise cooling or warming effects on the margins of the land-masses near which they flow if onshore winds blow over them. Of more local importance are the presence of large lakes, the influence of *physical features* involving differences in aspect, shelter from or exposure to cold winds, the presence of deep valleys and basins, and the influence of *soils* and *vegetation*.

THE ATMOSPHERE

The atmosphere is a thin layer of gas held to the earth by gravitational attraction, three-quarters of which lies within 11 km (7 miles) of the earth's surface, 90 per cent within 16 km (10 miles) and as much as 97 per cent within 27 km (17 miles). The 'weathermaking' layers are limited to a height of a few km, particularly as a large proportion of the water-vapour is contained in the lowest 3000 m (10,000 ft); more than half, in fact, is below 2300 m (7500 ft).

In recent years much research has been carried on into the physical conditions of the upper air. The *balloons-sondes*, carrying self-recording instruments, have been gradually replaced by *radiosondes*: radio transmitters, giving out signals indicating changes in pressure, temperature and humidity, are borne by balloons to a height of 18,000-30,000 m (60,000-100,000 ft), when they burst. The rawinsonde (an abbreviation for 'radar wind-sounding') carries a radar target enabling the -course of the balloon to be followed directly and so the upper air currents to be measured. More recently much information has been obtained by means of rockets and orbiting satellites. Observations thus derived may be plotted in various ways. Their interpretation can reveal much information about the vertical structure of the atmosphere, and it is now known that it is zoned or layered in various ways in terms of temperature change, chemical composition and electro-magnetic properties; a formidable jargon of -spheres has been coined.

It has been known for a long time that the temperature of the lower atmosphere falls at an average rate of 0-6°C per 100 m (3-3°F per 1000 ft) of ascent; this is termed the *environmental* (or static) *lapse-rate*. It is now realized that this decrease of temperature is limited to a certain height, which seems to be about 16 km (10 miles) at the Equator, 11 km (7 miles) at latitude 50°, and probably 9 km (5-5 miles) at the Poles, although these figures vary with the season (the point of change seems to be higher in summer) and with general atmospheric conditions. The lower atmosphere can be divided into two parts—the troposphere up to this level of change, the stratosphere up to about 50 km (30 miles); the discontinuity plane between them is known as the tropopause. At the latter, temperatures over the Equator vary during the year only from about –79°C (– 110°F) to –90°C (–130°F), though over the polar regions the seasonal difference is more marked, from about –40°C (–40°F) in summer to –79°C (–110°F) in winter.

Until recently the stratosphere was regarded as extending upwards to about 80-90 km (50-55 miles). It was known that from the tropopause to about 30 km (18 miles) temperatures rose very slightly. But still higher a change was found to occur; temperatures rose steeply to about 77°C (170°F) at 50 km (30 miles) and then fell equally steeply to about –100°C (–150° F) at 80 km (50 miles). This zone, therefore, from 50 km (a level known as the stratopause) to 80 km (the mesopause), which was formerly regarded as the upper stratosphere, is now called the *mesophere*. Within the stratosphere with a maximum at 25 km) is a zone of ozone concentration, which causes the absorption of the shorter ultra-violet rays from the sun's radiant energy, with the resultant heating of the thin atmosphere and the formation of this 'hot layer'. The longer ultra-violet rays, however, can pass through the ozonosphere to the earth's surface, producing the bronzing or 'sun-tan' which so many people desire.

Near the mesopause is the region of the silvery white noctilucent clouds, clearly visible in high latitudes during summer nights. It is not known whether these clouds are of water-vapour, ice-particles, or volcanic or ice-coated meteoritic dust; when a giant meteorite fell in Siberia in 1908, an unusually brilliant display of these clouds was seen the same evening, which seems to confirm that they are of dust.

Above the mesopause extends a zone known as the thermosphere, where temperatures would reach enormous levels, but since the density of the air is so small as to be a near-vacuum, very little heat can be absorbed, held or conducted.

The thermosphere broadly corresponds with the ionosphere, the zone in which electro-magnetic waves, including radio signals, are reflected back to earth. In this zone the atmospheric gases are ionized by incoming solar radiation. Above this again is the magnetosphere, the existence of which was postulated after two satellites, *Explorer I* and *II*, sent back to earth evidence of two zones of intense radiation, known as the Van Alien belts, the outer extending to about 16,000 km (10,000 miles) from the earth.

In these upper zones can be observed the Aurorae Boreah's and Australis in their respective hemispheres; though best developed at about 100 km (60 miles) altitude, they have ben recorded as far as 1000 km (600 miles). Their luminous effects are visible in high latitudes as red, green and white arcs, draperies, streamers, rays and sheets in the night sky. Probably this effect is the result of magnetic storms and of electrical discharges from the sun during periods of sun-spot activity or 'flares', mainly of electron particles funnelled into the earth's magnetic field and accelerated to the high energies necessary, thus causing ionization of gases. This seems to be con- firmed by recent American experiments, when an artificial aurora was produced by a rocket-borne device which fired bursts of electrons into the atmosphere. Results of the aurora include changes in the earth's magnetic field and black-outs in radio communication.

Constituents of the Atmosphere

Air is a mixture of gases, consisting mainly of about 78 per cent nitrogen and 21 per cent oxygen. Small quantities of carbon dioxide, hydrogen, and argon and other inert gases make up the remaining 1 per cent. These gases play a vital part in the radiation balance of the earth-atmosphere system.

In addition, air contains a variable quantity of water-vapour which plays a major role in the various weather phenomena. There is also much dust, specks of carbon in the form of soot, the spores of plants, salt particles evaporated from spray blown from the surface of the ocean, and minute pieces of disintegrating meteors, known as cosmic dust. One function of these particles is to serve as nuclei for the condensation of water-vapour.

It is now known as the result of space research that the constant chemical composition of the atmosphere refers only to that portion extending to a height of about 80 to 90 km (50 to 55 miles), to which has been given the name homosphere, extending broadly to the top of the mesosphere. Above the homosphere is the heterosphere, which consists of a series of concentric layers, in turn of molecular nitrogen, oxygen, helium and hydrogen. Though, interesting, the heterosphere obviously has little significance for the meteorologist.

TUNDRA CLIMATE

Location

Tundra is a Finnish word which means barren land. Thus, the tundra region having least vegetation and polar or arctic climate is found in North America and Eurasia between the southern limit of the permanent ice caps in the north and the northern limit of the taiga or subarcitc climate in the south. Thus, tundra climate has developed over parts of Alaska (USA), extreme northern parts of Canada, the coastal strips of Greenland and the arctic seasboard regions of European Russia and northern Siberia. Besides, tundra climate has also developed over arctic islands (Fig. 8.8). Vegetations rapidly change to the north of treeline because of increasing severity of climate.

Tundra climate is further divided into two subtypes *e.g.* (i) *arctic tundra climate* and (ii) *alpine tundra climate* (which is found over high mountains of tropical to temperate areas). Based on variations in general vegetation characteristics arctic tundra is divided in 3 zones from south to north viz. (i) low arctic tundra, (ii) middle arctic tundra, and (iii) high arctic tundra. It may be pointed out that high, middle, and low are not indicative of altitudes rather these indicate latitudes.

The poleward boundary of tundra climate is demarcated by 0°C isotherm of the wannest month of the year while 10°C isotherm of the warmest month makes the equatorward boundary.

The tundra climate is characterized by general absence of insolation and sunlight and very low temperature throughout the year. The average annual temperature is –12°C. Winters are long, bitterly cold and very severe while summers are very short but cool. The warmest month of the year records average temperature between 0°C and 10°C. It is interesting to note that diurnal range of temperature is very low because of very little difference in day and night temperatures but the annual range is quite large. The severe climate does not favour much vegetation growth and hence most of the areas under tundra climate remain barren land. The ground surface is covered with snow at least for 7 to 8 months each year. The region is swept by speedy cold powdery storms known as blizzards. Growing season is less than 50 days in a year. The ground is permanently frozen (permafrost). Even soil is also perennially frozen.

Precipitation

Mean annual precipitation, mostly in the form of snowfall, is below 40 cm. The absolute humidity is very low because of very low rate of evaporation due to very low temperature throughout the year. The divergent system of air circulation and anticyclonic conditions do not favour much precipitation. Most of the annual precipitation is received during summer and autumn because of relatively higher temperature.

Natural Vegetation

There is prefect relationship between vegetation and the condition of moisture in the soils. The characteristic lithosols of tundra biome support only lichens and mosses. Only 3 per cent species of the total world species of plant could develop in tundra climate because of the severity of cold and the absence of minimum amount of insolation and sunlight. The vegetations of tundra climate are cryophytes *i.e.,* such vegetations are well adapted to severe cold conditions as they have developed such unique features which enable them to withstand extreme cold conditions. Most of the plants are tufted in form and range in height between 5 cm and 8 cm. These plants have the tendency of sticking to the ground surface because the temperature of the ground surface is relatively higher than the temperature of the overlying air. The herbs have developed only in those areas where heaps of ice and snow protect the plants from gusty icy winds. Such herbaceous plants include willow the stems of which are very close to the ground surface (hardly a few centimetres above the ground). Though the growth rate of these herbaceous plants is exceedingly slow but their survival period is unbelievably very long (between 150 to 300 years). The evergreen flowering plants develop on the ground like cushions mostly during short cool summers.

Index

B

C

L

M

N

O

P

□□□